Praise for *Purr*

"A wonderful, evidence-based guide with concrete suggestions for cat owners to improve the lives of the cats they live with. *Purr* is a welcome and exciting addition for every cat lover's bookshelf."

MIKEL DELGADO, PhD, certified applied animal behaviorist and cofounder of Feline Minds

"What a gift to applied animal behavior is Zazie Todd. *Purr* is both packed with the latest findings and eminently readable. I could not recommend it more highly."

JEAN DONALDSON, the Academy for Dog Trainers

"*Purr* is an absolute must-read for all animal lovers! This fantastic, evidence-based guide is bound to influence cat-guardian relationships for the better."

SARAH-ELIZABETH BYOSIERE, director of the Thinking Dog Center and assistant professor at Hunter College (CUNY)

"Practical, entertaining, and thorough. Cat owners are in good hands with Zazie Todd."

LUCY JANE SANTOS, author of *Half Lives: The Unlikely History of Radium*

"Cat lovers learn the science behind cats' petting preferences; the multiple meanings of purrs, chirrups, and meows; how to best satisfy the scratching and stalking desires for indoor cats; and even how to keep both cats and wildlife safe if your felines spend time outside."

CAT WARREN, *New York Times* best-selling author of *What the Dog Knows*

"*Purr* is cat-centered, so much that it feels like a very smart cat wrote it! Starting with the most recent scientific findings, Zazie Todd explains with clarity what cats enjoy, what is good for their well-being, and how we can build a strong reciprocal bond with them. Indispensable reading."

CARLO SIRACUSA, coeditor of *Decoding Your Cat* and board-certified veterinary behaviorist, UPenn School of Veterinary Medicine

Purr

THE SCIENCE
OF MAKING YOUR
CAT HAPPY

ZAZIE TODD

Foreword by **PAM JOHNSON-BENNETT**

GREYSTONE BOOKS
Vancouver/Berkeley/London

Greystone Books Ltd.
greystonebooks.com

Cataloguing data available from Library and Archives Canada
ISBN 978-1-77164-814-1 (cloth)
ISBN 978-1-77164-810-3 (epub)

Editing by Lucy Kenward
Copyediting by Rowena Rae
Proofreading by Meg Yamamoto
Indexing by Stephen Ullstrom
Jacket design by Belle Wuthrich
Text design by Fiona Siu
Jacket photograph by iStock/spxChrome

Printed and bound in Canada on FSC® certified paper at Friesens.
The FSC® label means that materials used for the product have been
responsibly sourced.

Greystone Books gratefully acknowledges the Musqueam, Squamish,
and Tsleil-Waututh peoples on whose land our Vancouver head office
is located.

Greystone Books thanks the Canada Council for the Arts, the British
Columbia Arts Council, the Province of British Columbia through
the Book Publishing Tax Credit, and the Government of Canada for
supporting our publishing activities.

Canada

For all the cats,
but especially Harley and Melina.

CONTENTS

FOREWORD

WHEN I BEGAN my career in cat behavior consulting in 1982, it was a lonely field. It seemed that animal professionals were more focused on dog training and health. Over the next four decades, the field has extended to our feline pets, I'm glad to say, and Dr. Zazie Todd has played a role in that welcome shift in both attitude and support.

Too many people dismiss cats as independent, untrainable, aloof, and unaffectionate. Through my own writing and whenever I speak or am interviewed, one of my goals is to change that impression. I wrote the book *Think Like a Cat* to help cat owners see life from a cat's point of view. It was a groundbreaking book, because it was one of the first to highlight the importance of a cat's emotional health. It wasn't just about training a cat to behave, but also about understanding why cats do what they do and how to strengthen the bond you share. Every behavior serves a purpose, and when you stop labeling actions as misbehaviors, you begin to enjoy how truly smart, affectionate, social, and tolerant cats are.

The cat behavior field continues to grow as more people appreciate the importance of cat happiness, enrichment, stress reduction, and positive training. That said, cat owners often don't know where to turn for accurate, science-based information that they can put into practice in daily life. Many still go by past beliefs about cats or follow advice of friends or what they read on the internet. This is where Dr. Zazie Todd's work is making a difference. Her training in psychology and her knowledge of animal training and welfare, paired with a deep love and understanding of how cats think, make her a terrific resource. In *Purr: The Science of Making Your Cat Happy*, she shares sound, evidence-based research and theory along with years of professional, hands-on experience. She makes the techniques easy for any cat owner to implement by ending each chapter with tips to apply the science at home. Dr. Todd gets to the heart of what cats need and, most importantly, why they need it.

I have been a fan of Dr. Todd's *Companion Animal Psychology* website for quite a while, and learned so much from reading her first book, *Wag: The Science of Making Your Dog Happy*. She does all the heavy lifting for us by tracking down and dissecting the latest research, and she then interprets and shares it in a way every cat owner can understand. Todd's love for cats shines through in her writing. She is also a natural teacher and one of the best spokespersons for our beloved companion animals.

Purr: The Science of Making Your Cat Happy offers valuable insights into that beautiful, graceful, and intelligent creature known as *your cat*. Dr. Todd takes you on an in-depth yet very accessible tour of what cats need physically, environmentally, and also emotionally. This is where science meets heart.

From kittenhood through your cat's golden years, happiness awaits you and your favorite feline within the pages of this marvelous guide.

—PAM JOHNSON-BENNETT, CCBC

INTRODUCTION

SOFT YET CLAWED, with a beautiful purr yet considered inscrutable, cats are much misunderstood. As with so many things in life, the more you put into your relationship with your cat, the more you will get back. Whether kept indoors or at least partly outdoors, as a single cat or in multi-cat homes, all cats have particular needs. After years of writing about and working with cats, I'm thrilled to write about how taking account of cats' needs can give you a better human-feline relationship and, along the way, to share the story of my own cats, Harley and Melina.

I earned my PhD in psychology at the University of Nottingham, where I taught small group classes in animal behavior and other introductory psychology topics, even dissecting sheep's brains (in the days before guidance on bovine spongiform encephalopathy meant this practice was inadvisable). I worked as a social psychologist for some years, conducting research on the perception and communication of risk, before emigrating to Canada with my husband and completing an MFA in creative writing at the University of British Columbia. After adopting two dogs and two cats in a short space of time, I was surprised by some of the information I came across, which didn't fit with what I already knew about animal behavior from my background in psychology. In particular, I was concerned about recommendations to use harsh training methods. So I began to do more research on what science tells us about the behavior of companion animals—and found the fascinating fields of canine and feline science.

In 2012, I started my blog, *Companion Animal Psychology*, to share quality information about cats and dogs with ordinary pet lovers. I found that there's a real need for articles that are evidence based. I was thrilled to win a scholarship to study at the Academy for Dog Trainers, known as the Harvard of dog training, where I learned efficient training methods and how to deal with behavior problems such as fear and aggression—the practical application of the psychological science I had taught so long ago. I also took an Advanced Certificate in Feline Behaviour from International Cat Care, a wonderful course that covers all aspects of how domestic cats experience the world and behave in it. Along the way, I learned a huge amount from volunteering with cats, dogs, and small animals at my local branch of the British Columbia Society for the Prevention of Cruelty to Animals (BC SPCA), a world-leading organization. I enjoy helping people resolve behavior issues with their dogs and cats through my business, Blue Mountain Animal Behaviour. And I'm lucky enough to still be involved in some research on companion animal behavior, and to teach a course on communicating anthrozoology for the master's students in the anthrozoology program at Canisius College (anthrozoology is the study of people's relationships with animals).

Even before I wrote my book *Wag: The Science of Making Your Dog Happy*, I knew that I would like to write a similar book about cats. People with cats have the same need and desire to learn more about our knowledge of cats and to have solid, evidence-based information on how to care for them. I hope this book will bring you lots of fascinating new knowledge about cats and help to transform your own relationship with your cat or cats.

This book is not a substitute for a professional opinion and cannot provide advice on your specific cat(s). If you have any concerns about your cat(s), seek advice from a veterinarian, veterinary behaviorist, and/or cat behavior expert as appropriate.

1

HAPPY CATS

·················

MELINA IS BUSY trying to distract me from writing this book. She has already walked in front of the computer screen two times, and just now she's come to sit next to my keyboard, looking hopeful. Of course I can't resist. I reach out a hand and she sniffs it and then rubs the side of her head on it. Next she raises her head slightly to make it easier for me to pet her under the chin, just how she likes it. Her purr is soft and melodic. Then, satisfied, she leaps up to the window ledge behind my monitor to watch the world outside for a while, the tip of her tail twitching ever so slightly.

Anyone who thinks that pet cats don't care about their people isn't paying enough attention. This is a problem for cats: people think they are easy pets so don't provide what they need. We have so many stereotypes about cats as loners and jerks and just difficult animals that it's as if no one sees the actual cat in front of them: a beautiful, fluffy bundle who wants and craves your attention (albeit on their own terms), who delights in chasing the wand toy, and who loves to snooze in the sunny spots of your home, but who also likes somewhere small and cozy to curl up in to relax and be safe. When you see cats for who they are and give them what they need, they will be happier, they will be less likely to have behavior problems, and your efforts will be repaid with feline affection.

For most cats, happiness isn't being squished and petted for half an hour and then ignored for the rest of the day. Cats have their little quirks—and that's what we love about them—but at the same time, every cat has a set of needs that we should meet. And I am sure they care about us. Take the way Melina just jumped back to my desk and sniffed noses with me (a feline greeting) before rubbing her head on my forehead, soft warm fur and no doubt some pheromones that I can't detect left behind in a streak against my skin. This head-rubbing, or bunting as it is technically called, is an important behavior between cats who are part of the same social group. Now Melina has leaped up to the top of the bookshelves, where she can survey the room but is still close by (and can watch my work). If I look up at her—there—I get a slow-blink, which, of course, I return.

Meanwhile my other cat, Harley, a hefty tabby, is under my desk near my feet. No doubt this has something to do with the fact that the heat is on, and the heat vent under my desk is blowing out nice warm air, but still, he has the whole house with a heat vent in every room to choose from. Indeed, he has already spent part of the morning, as he usually does, completely blocking the heat vent in the hall so that he is nice and warm but no hot air escapes around him into the house. And now he's picked the vent near me. In his own way, he is choosing to be close to me. And probably soon, like most days, he will jump on my lap and demand lots of petting, then climb on my desk, trample my keyboard, and get picked up and removed to my old office chair, which I have had to keep, because he likes to relax in it so much.

The domestic cat is descended from *Felis lybica*, a desert cat that is still found in parts of Africa. Occasionally, bones of cats have been found in Bronze Age and Iron Age settlements, and the presence of skinned cat bones in an archaeological site at Coppergate, York, shows that in Anglo-Saxon England at least some cats were used for their fur.[1] A ninth-century poem, "Pangur Bán," shows that sometimes cats had names, which speaks of cats as companions. But the oldest known case of a domestic cat living alongside humans dates from the Middle Ages in Kazakhstan, along the Silk Route by which people and goods moved from East Asia to Persia, East Africa, and Southern Europe.[2] Up until very recently indeed, domestic cats were prized for their abilities to catch rodents. Now, they are increasingly seen as part of the family and often live indoors, no mice in sight.

Everyone who loves cats knows that every cat has their own individual personality and preferences. When we think of pet cats in their homes, we think of much-loved pets who get to snooze all day long. But what if many of those cats are actually bored out of their mind and stressed by their home life?

WHAT CATS WANT AND NEED

LUCKILY, WE'RE NOT starting from scratch when considering what cats need. Several sets of guidelines on animal welfare can help tell us the best ways to care for pet cats. One of these, which has been around for decades now, is called the Five Freedoms. These were originally developed in the UK for farm animals but were then understood to apply to all animals, including companion animals. In many places, the laws around animal welfare and animal cruelty have some basis in the Five Freedoms, and failure to meet them can result in prosecutions for animal cruelty.

The Five Freedoms

- Freedom from hunger and thirst, by ready access to water and a diet to maintain health and vigor.
- Freedom from discomfort, by providing an appropriate environment.
- Freedom from pain, injury, and disease, by prevention or rapid diagnosis and treatment.
- Freedom from fear and distress, by ensuring conditions and treatment that avoid mental suffering.
- Freedom to express normal behaviors, by providing sufficient space, proper facilities, and appropriate company of the animal's own kind.

If you walk into an animal shelter, chances are you will see the Five Freedoms on a poster on the wall or find them on the shelter's website. They've made a tremendous contribution to animal welfare around the world. And so, we know that our pets have five welfare needs—diet, environment, health, companionship, and behavior.

In large part, the Five Freedoms are about preventing cruelty. Four of them are "freedom from" while only one is written positively, "freedom to express (most) normal behaviors." But as we've come to learn more about animals, we want more and better things for them. And we've learned so much about animals, including cats. In the past, scientists used to think that animals did not really experience emotions. Certainly we can understand that it was then—and still is now—hard to demonstrate beyond doubt that animals can have emotions, although it does seem that a belief that humans are special may also have contributed to this view. These days, thanks to pioneering work by many different scientists, we know for sure that animals experience emotions even if we don't know exactly how to describe their subjective experience. We can't ask a cat to tell us how they feel when basking in warm sunlight or watching hummingbirds fly to and fro from a feeder. But we can be sure that they are feeling something.

Scientists' work on animal emotions covers a wide variety of species and uses a range of methods, including neuroscience and experimental work. And researchers have used some inventive approaches to increase our knowledge. For example, we now know that fish feel pain.[3] They have a type of receptor called a nocireceptor that detects pain in response to stimuli, so that's one sign they can feel pain, but there's experimental work too. In one study, scientists observed what happened when they dropped something aversive that rainbow trout would normally avoid into a tank (they used a Lego brick) and compared how the trout responded in different conditions. When they were injected with acetic acid, which is painful, these fish did not avoid the bricks; it seems the pain was too distracting. But if the trout were injected with both acetic acid and morphine, a painkiller, they did avoid the bricks.

Another example, this time with lab rats: scientists taught lab rats to play hide-and-seek with them.[4] The rat was put in a box in a room, and the scientist closed the lid to indicate that the rat was to be the seeker; then the scientist hid in one of three designated places in the room and used a remote control to open the box. When the rat found them, they were rewarded with play. In the hide condition, the box stayed open, the scientist crouched down next to it to indicate what would happen, and the rat could jump out and choose from one of seven hiding spots. Each time, the rat was rewarded with play. The rats tended to pick an opaque (cardboard) box to hide in rather than a transparent one, suggesting that they knew when they were visible, and they were mostly quiet when hiding. But when seeking, they made excited noises (out of human hearing range but detected by the lab equipment). Sadly, as with most research on laboratory rodents, the rats were euthanized when the study was over. But studies like this leave us with no doubt that all kinds of animals feel *something*.

Many strands of research led scientists to issue the Cambridge Declaration on Consciousness in 2012.[5] In part, the declaration reads: "Convergent evidence indicates that nonhuman animals have the neuroanatomical, neurochemical, and neurophysiological substrates of conscious states along with the capacity to exhibit intentional behaviors. Consequently, the weight of evidence indicates that humans are not unique in possessing the neurological substrates that generate consciousness. Nonhuman animals, including all mammals and birds, and many other creatures, including octopuses, also possess these neurological substrates." And of course, this applies to the cat purring on your lap right now just as much as it applies to octopuses and birds.

Cats are sentient creatures whose behavior can be intentional and who experience a range of emotions.[6] Scientists have also been learning quite a bit about the cognitive abilities of cats.[7] Cats understand that things still exist when they go out of sight (object permanence), as demonstrated in experiments in which a piece of food is hidden behind a box and that's where the cat goes first to look for it (even if the sneaky experimenter actually hid it somewhere else). Scientists have trained cats to distinguish between two dots and three dots, showing that they have some concept of quantity (perhaps useful if they want to get a bigger meal). Cats have been shown to follow a point to find hidden food (although

I'm not sure anyone told Harley this!). And cats who have a good bond with their owner will look to their person for information when they see a potentially scary object (see chapter 8). In other words, cats are intelligent creatures with complex social worlds who will do much better if we provide more for them.

Now that we know more about the many and varied abilities of animals, it changes how we think about animal welfare. The Five Domains model updates the Five Freedoms to include positive experiences.[8] The first four of these domains will be familiar from what you've just read: good nutrition, a good physical environment, good health, and good behavioral interactions that allow the animal to express normal behaviors (with the environment, with other members of the same species, and with humans). "The fifth one is an emphatic commitment to giving animals opportunities to have more positive experiences," said David Mellor, the now-retired professor from Massey University in New Zealand who came up with this model. "Good nutrition, good environment, good health, appropriate behavior, and opportunities to have more positive experiences" are what's important for good animal welfare. He told me that "we need to make a distinction between what we need in order to get animals to survive, and what we need in order not just to have them survive but to have them thrive." I think all pet guardians can get behind the idea of helping our pets thrive.

The thing is, we can't completely prevent negative experiences. Some of those experiences are built-in biological mechanisms that help us survive. For example, if we never felt thirst, said Dr. Mellor, we wouldn't be motivated to drink; we drink to quench that sensation. These kinds of sensations—thirst, hunger, breathlessness, pain, etc.—are specific to particular experiences and motivate animals (including humans) to do specific things to solve the problem. They have evolved over time and are essential for survival. These experiences are largely covered by the first three of the five domains (nutrition, physical environment, and health). "Thirst promotes you to drink water; hunger, to eat energy-rich nutrients; nausea, to avoid what you've eaten; pain, to escape from or to avoid injury-producing experiences and events and stuff like that," he said. So what we need to avoid when caring for our animals is excessive thirst, which we can do by making sure water is always available. This doesn't create a positive state—lack of thirst is a neutral experience, not a positive one—but we can make these experiences positive. One example he gave me that definitely applies to cats is providing a source of radiant heat when it's cold; I just have to think of Harley hanging out on the heat vent to know that he enjoys the feeling of warmth.

Although this kind of experience (thirst and so on) is created internally by the animal, it's important to consider another type of experience that can have big implications for animal welfare, and that we also have a lot of power to change: experiences due to how they perceive the environment. "They're the ones like anxiety, fear, loneliness, boredom, rage, anger, depression, things like that," said Dr. Mellor. "These are negative experiences of that character, generated by the animal's perception of its external circumstances." The situation in which cats live is largely (or entirely, for indoor cats) under human control. Here, there are many ways to provide positive experiences: letting them explore the environment, for example, and encouraging them to simulate hunting with wand toys and other cat toys.

Interactions with us and with other animals in the home can provide positive social experiences. One way to tell if cats are enjoying these experiences is simply to look at whether they are taking part in them. Are they playing with their new toy, for example? Unfortunately, a cat that is in pain or afraid will not feel like taking these positive opportunities. That's why it's so important to limit negative experiences as much as we can, because they not only are unpleasant in themselves but also stop the cat from doing positive things.

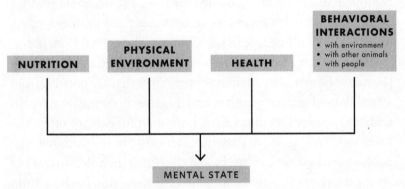

The four physical domains all contribute to a cat's mental state, which in turn contributes to their overall welfare. Based on Mellor (2016) and Mellor et al. (2020).[9]

Opportunities to express normal behaviors may require the presence of another animal of the same species. Being housed with (or apart from) other animals is considered an important feature of animal welfare, and for companion animals, there is now recognition that interactions with people may be an important component of good welfare too. Clearly, interactions that are scary or frightening for the animal may be a threat to their welfare. Some cats (particularly those who were not socialized with people as kittens) will prefer not to interact with people, or to have only one or two people they interact with on a regular basis. But for cats who were socialized as kittens and who enjoy being around people, positive interactions with people are an important part of their welfare (and are also mentioned in the five pillars of a healthy feline environment, which we'll get to in chapter 3). When cats have a good relationship with people, they choose to approach them, want to spend time with them, and show relaxed and positive body language in their presence.[10] People can be good for animal welfare when they provide companionship, train with positive reinforcement, provide nice food and desirable petting, engage the animal in enjoyable activities, act as a source of safety (a secure base: see chapter 8), and protect the animal from harm.

In 2020, the Five Domains model was updated to incorporate the role of humans in providing experiences that are positive from an animal welfare perspective, and this is another way in which it updates the Five Freedoms.[11] It's important for cats (or other animals) to have a choice in these interactions. The more knowledge people have of what cats need, and the better their skill in dealing with cats, the better those human-cat interactions will be from the cat's perspective. In contrast, when people are not skilled, use aversive training methods (such as squirting with water), do not give the cat a choice, leave the cat home alone for too long, don't take the cat to the vet when needed, and so on, the cat's welfare will suffer.

The first four domains—nutrition, physical environment, health, and behavioral interactions—all feed into the fifth domain, the animal's mental state. For cats, alongside the five pillars, the Five Domains model is a helpful framework to consider how we should best care for them, and that leads us to consider not just what pet cats need as a species but also what individual cats need and like. When we don't provide those things, our cat's welfare suffers. So what are the biggest threats to feline welfare?

THE MAIN THREATS TO CATS' WELFARE

EVERY YEAR, AN animal charity in the UK called the People's Dispensary for Sick Animals (PDSA) publishes its PDSA *Animal Wellbeing (PAW) Report* that assesses how well British pet owners are meeting their pets' needs.[12] The report covers dogs, cats, and rabbits and uses a representative sample, so the results can be generalized to the UK as a whole. Although other countries may differ, this report gives us a snapshot of where we're at in terms of animal welfare.

The 2011 *PAW Report* described pets as "Stressed. Lonely. Overweight. Bored. Aggressive. Misunderstood ... but loved." Amongst positive changes for cats since then, the 2020 *PAW Report* shows a big increase in the proportion of cats getting micro-chipped, which is an important way to reunite animals with their owners should they get lost, and now at 74 percent of British cats. But overall, there seem to be more downsides: fewer cats receiving all of their vaccinations when young, down to 69 percent; only 84 percent of cats being registered with a vet, unchanged since 2011; an increase in the number of cats being kept indoors only, even though some outdoor access is the norm in the UK (see chapter 4 for more on this subject); and people being less informed about pets' need to be housed with (or apart from) other animals and about the need for opportunities to express normal behaviors.

Every year, the *PAW Report* questions ask about people's knowledge of pets' welfare needs. Those three needs that are critical to survival—good nutrition, a good environment, and protection from pain and suffering—are the most well known. Eighty-six percent of pet owners know about the need to express normal behaviors, and 81 percent identify the need to be housed with or apart from other animals. But until they took part in the study, only a fifth of people knew that pets have five welfare needs.

Cats are a flexible species when it comes to social relationships. Some cats get on well with other cats, especially those who grew up together and/or who were well socialized during the sensitive period for socialization when they were kittens (although there are no guarantees). But cats are solitary hunters, and so they don't need to live with another cat in order to survive. In fact, many cats prefer not to live with other cats. This means each cat's need for companionship or otherwise has to be evaluated individually. Some cats are very stressed at being forced to live in a household with other cats, or even near other households with cats. Although the PAW Report found that most people say their cats get on, up to a fifth of cats are living with another cat they do not like. That makes providing resources for each cat especially important in multi-cat households.

But people were not providing extra resources to reflect the number of cats they had, with the exception of food bowls. Amongst households with more than one cat, 64 percent of cat owners provide either just one or no litter boxes (a general rule of thumb is to provide one litter box per cat plus one spare). If cats don't have enough resources, they can become stressed. Seventy-seven percent of cat owners said they would like to change at least one of their cat's behaviors, most commonly scratching furniture or carpets, waking people up in the morning, or begging for food.

Unfortunately, we may be the cause of some of these behavior problems. We assume that much-loved cats in homes are well cared for, so it comes as a surprise to many that cats in this environment may be bored, stressed, and overweight. A poor home environment can cause stress that in turn leads to behavior issues. In fact, this is the main welfare issue affecting pet cats, according to research published in *Veterinary Record*.[13] Cat experts were asked to name and discuss the most important welfare issues for domestic cats, and then to rank them according to severity, duration, and prevalence. The results show that for individual cats, social behavior issues (those problems people complain about) because of a poor home environment are the most significant threat to their welfare. This is because of the severity of the problem and the fact it can go on for a long time. If cats don't have what they need, the anxiety and stress can contribute to a range of behavior issues including house soiling. Ultimately, if unresolved, these behaviors can result in people giving their cat up to a shelter.

A poor environment for the cat can simply mean it is not set up right for them. For example, cats need hiding places and opportunities to express normal behaviors (such as scratching and play). To learn more about cats' environmental needs, check out the five pillars of a healthy environment for cats (chapter 3). The scientists point out that cats living in groups are not necessarily more stressed than singleton cats, but competing with other cats for access to things like litter boxes can result in stress and inter-cat aggression. However many cats you have, it is important to ensure that all cats have access to what they need, and it's also a good idea to give them lots of enrichment—especially if they are indoors only. Keeping cats indoors means that any deficiencies in their home environment are exacerbated, because the cat has nowhere else to go.

The second most serious welfare issue for individual cats, according to this study, is diseases of old age, which include dental disease, arthritis, diabetes, and cognitive dysfunction (see chapter 12). The scientists say that both owners and veterinarians will sometimes put changing behaviors or health down to old age, which means cats may not be taken to the vet and may not receive treatment for potentially treatable issues. The result for cats is pain or ill health. Obesity is the third most serious welfare issue for individual cats, according to this research: it is estimated that up to 45 percent of cats are overweight or obese, and this condition is linked to diabetes, cardiovascular disease, urinary tract disease, and orthopedic disorders. People need to recognize (or be told by their vet) when their pet is overweight and help their cat lose weight. Another serious welfare issue for individual cats is not taking them to the vet when they need to go. Many owners struggle with getting their cat in the carrier or don't realize when their cat needs to see a vet. Other welfare issues mentioned are poor management of pain and issues with shelters and unowned cats.

The same study also asked cat experts to rank welfare issues in terms of how common they believe them to be. The most common welfare issues are neglect and hoarding. Hoarders have more cats than they're able to take proper care of, although they're often in denial about this reality and believe they're helping the cats. Hoarding is a complex issue that requires a multi-agency approach to resolve. Other common welfare issues include delayed euthanasia when the cat's quality of life is poor, inherited diseases (including issues to do with conformation; e.g., brachycephalic cats with squashed faces), restriction of the environment leading to behavior problems, and poor pain management.

Prof. Cathy Dwyer of Scotland's Rural College, coauthor of this research, told me in an email, "I would most want cat owners to understand more about cat behavior—why cats do what they do, what they need for good welfare and how we can provide that for them. To be honest (and as an ethologist I might just be a wee bit biased here), if we paid more attention to animal behavior then a lot of the other issues (relinquishment to shelters for behavioral issues, not recognizing pain, not seeking veterinary treatment, etc., etc.) might just go away as well!"

A study published in *Animals* provides support for the idea that better knowledge of cats would lead to happier cats with fewer behavior issues.[14] Some of the positive results from this study include that most people give their cats toys that they can use to play by themselves (e.g., toy mice); most people scoop the litter box at least once a day; almost a third of people have taken the time to train their cat to do something, such as to play fetch or do a trick; and almost half of cat owners take time to play with their cat every day. Less good is that if they saw their cat do something to misbehave, 85 percent of people said they made a loud noise, 77 percent said they would yell, 51 percent had tried spraying the cat with water, and 11 percent had tried kicking or hitting the cat. People who were more knowledgeable about cat behavior were less likely to use these techniques and also less likely to have a cat with a behavior problem. They were also more tolerant of most behavior issues (albeit not of house soiling).

Also of note: in this study, 92 percent of people said their cat was part of the family. This is comparable to a survey in Victoria, Australia, that found 89 percent agreed their cat was part of the family.[15] In that study, most people were doing a good job of meeting their cat's welfare needs, but again there were some places where improvements could be made. It seemed likely that some people did not realize when their cat was overweight; 11 percent of homes did not provide toys for their cat; quite a few homes did not have enough litter boxes; and a quarter of cats had not been to the vet in the last year (and 6 percent had *never* been to the vet). Dr. Tiffani Howell, first author of the study, told me, "Cat owners appear to be meeting their cat's welfare needs, with a few areas for improvement. For instance, nearly half of owners allow their cat to roam free outdoors, which could lead to injuries. Female owners report higher levels of satisfaction with their cat's behavior, and fewer behavioral problems, than male owners. Older owners were less likely to have irretrievably lost a cat than younger owners, but they report more behavioral problems."

Making sure your cat is happy doesn't mean buying lots of things for them. I only have to watch Melina play football with a dried edamame bean or a curled-up ball of paper to know how much pleasure cats can find in very inexpensive things. Making your cat happy is about finding out what matters to them, and is most likely more to do with how you allocate your time rather than how much money you spend. Veterinarian Dr. Kat Littlewood, PhD, who is a lecturer in animal welfare at the School of Veterinary Science (Tāwharau Ora) at Massey University in New Zealand, says that, especially when talking about indoor cats, people "have to be really aware that they need to provide for agency for their cat. So they need to provide behavioral enrichment or environmental enrichment, whatever's going to replace the fact that their cat isn't going outdoors. I can't over-emphasize the fact that we don't want to just think about alleviating negative things. Like we can give them heaps of food, we can make them really comfortable, we can buy lots of fancy cat beds for them, but if that's actually not what's important to the cat, if they're not actually happy and enjoying life, then, I dunno ... There's more to a cat's life than just being fed and safe. They also need to be happy."

For cats to have a better life, it would be helpful if everyone—not just cat owners—understood more about cats. This would change the stereotype of cats being easy pets. Veterinary behaviorist Dr. Karen van Haaften of the BC SPCA says, "Just as people need to accept that if you're adopting a dog you need to set aside an hour a day at least to walk that dog, you also need to put aside that time to play with your cat on a daily basis. They have needs in the same way that dogs do. Basically I think a lot of people think of cats as a kind of lower-maintenance pet and I don't think that's necessarily fair if you're going to give that cat the enriched life that they deserve. They require work and commitment and consideration of all of their needs."

As well as providing enrichment, making sure your cat is happy means paying attention to behavior issues, because it's important to find out if there are medical or environmental causes to fix. Many people with cats that display signs of behavior problems never discuss those issues with a vet or do anything about them.[16] This finding shows the importance of teaching people what cats need in order to have good welfare, and what can be done to help with behavior issues.

Part of the difficulty with understanding cats is that scientists don't yet know as much about them as we might like, and even what they do know is largely not accessible to cat guardians. Dr. Lauren Finka, feline scientist at Nottingham Trent University in the UK and author of *The Cat Personality Test*, told me,

"In terms of society, we've had them [cats] as pets for a very long time. We very much integrate them into our homes, but at the same time we don't necessarily understand a lot about their species-specific needs. We still don't know a lot about their behavior and body language. This is primarily because they're a very hard species to study and to work with, and there hasn't been a lot of research in this area compared to dogs, so that makes things difficult. And also most people aren't engaged with the scientific literature, so it can be very difficult for the average person to really understand in-depth about their cat's behavior and their expressions."

The aim of this book is to bring some of that science to cat owners and to show how it can help us not only to understand our cats better but also to be better cat guardians. This book will help you learn what your cat needs and identify what your cat likes. To do so, it's important to learn to tell when your cat is happy and when they are not.

"One thing that would make a huge difference in the lives of cats is if we could finally change the misconceptions people have about them. Hatred of cats, punishment, neglect and callous attitudes develop because people believe myths and false information about what drives cat behavior. Cats are often viewed as either sinister bird killers or low maintenance alternatives when you don't want to put much effort into being a pet parent. Education can open hearts to how wonderful it is to love and be loved by these magnificent animals."

—**PAM JOHNSON-BENNETT**, best-selling author of books including *Think Like a Cat* and star of TV's *Psycho Kitty*

HOW TO TELL IF YOUR CAT IS HAPPY

RESEARCH ACROSS A range of species has shown that animals experience emotions, but the work of the late neuroscientist Jaak Panksepp has been especially influential. You may have heard of him because of his work on tickling rats. Panksepp identified seven primary emotion systems in animals' brains, some of which are positive and some negative.[17] On the positive side, we have SEEKING, which includes curiosity, play, and anticipation (all of these emotional systems are written in capitals to distinguish them from emotional experiences). PLAY, LUST, and CARE (e.g., of young) are the other positive systems. Meanwhile on the negative side, there are RAGE (anger), FEAR (which is self-explanatory), and PANIC (sadness or loneliness). Panksepp's research was one of the important contributions that led to the Cambridge Declaration on Consciousness.

When you look at your kitty cat, you don't know what's happening inside their brain. But it is really helpful to have a good idea of how to recognize feline emotions from their behavior. For ease of explanation, let's consider feline body language in two groups: signals aimed at decreasing distance (purr purr, come closer) and signals that mean the cat is trying to increase distance. Pay attention to all of your cat's body parts when trying to understand their body language, and to the speed at which actions happen. For example, a slow-blink is a sign of a relaxed cat (and it's a good idea to slow-blink back—see chapter 8), whereas a rapid blink with scrunched-up eyes is a sign of fear.[18]

Allie relaxes on a bed. *JEAN BALLARD*

Essentially, the more open your cat's body position, the less stressed they are. A cat who is lying on their side or their back with their legs stretched out, their tail loose and out, and their belly on show is relaxed. Their eyes might be fully or partially closed, or doing a slow-blink at you if you're lucky, and the ears and whiskers will be in a normal position. Signs that your cat wants you to stay away include a swishing tail, rippling skin, a flattened body (as if trying to be invisible), the limbs and tail tucked in close to the body, hissing, spitting, and caterwauling. If you see these signs, don't approach your cat, because they may feel forced to defend themselves with tooth and/or claw. Instead, give them space. If you were petting them, stop—before they have to make it even more clear that enough is enough. It is always best to let a cat approach you and give them a choice of whether or not to interact with you.

Cats can become frustrated because they are bored, unable to access something they want (such as food or a bird they can see through a window), or unhappy at being in a shelter, where they may make a mess of their cage by tipping out their food and water bowls, knocking over their litter tray, and attempting to escape by putting their paws through the bars, pushing their body against the cage door, and biting the cage.[19] Signs of frustration include this kind of destruction and trying to escape, as well as meowing a lot, pacing up and down, and rubbing on things a lot. They will be highly aroused with a thrashing tail, and you may see their skin ripple too. Be very cautious with a cat in this state, as you may be at risk of a redirected bite—one where you aren't the cause of the aggression, but you happen to be there so you get the result of it. You can help to prevent frustration by figuring out what the cat wants and finding ways to safely provide it to them.

Humans often look at facial expressions for clues about emotion, and yet the fact that there's such wide variety in the shapes of cats' faces makes it hard to learn to read their emotions this way. But scientists are making advances in this tricky area. In one study, they looked at the faces of cats in shelters both during petting (with a hand or a touch stick) and while just in the cage.[20] Because of the different face shapes, the researchers looked at how the face changed from a normal position. They were especially interested in what might be signs of FEAR, SEEKING, and RAGE. Being afraid was associated with a low, flat posture, with hiding, and with retreating and moving back, often with a freeze (a moment of keeping still) beforehand. As well, they found that the direction of the cat's head was associated with their emotions; a leftward gaze or turn of the head when the cat was fearful, and a gaze or turn to the right in a relaxed cat. Some of the cats in the study were frustrated, and signs included licking the nose, the tongue being visible, and hissing.

Scientists have also developed and validated a feline grimace scale that may be useful to veterinarians wondering if a cat is in pain or not.[21] The facial signs they identified are a change in the muzzle shape (oval and bulged rather than round); changes in the position of the tips of the ears (far apart and facing out); a narrowing of the eye area with the eyes potentially partly closed; and the whiskers being forward, straight, and away from the face, instead of the usual lovely, relaxed curve. As well, the head was lower than usual, below the shoulders or with the chin tucked close to the chest. This scale is designed for professional use and these signs may not be easy to spot, but if you are ever concerned about your cat's health, always see your veterinarian.

Cats' reputation for being hard to read is part of their mystique. And we love them almost because of this, rather than despite it. If you find it hard to read your cat's body language, you are not alone. Most people performed worse than chance when scientists showed them videos of cats and asked them to say if the felines were feeling positive or negative emotions (happy or unhappy, if you like).[22] Only 13 percent of people were good at it, and they weren't your average pet owner—rather, they were people who worked with cats, such as veterinarians, vet techs, and people who work in animal shelters. Experience with many cats may play a role in discerning these emotional shifts, but it could also be that for our own cats in our own homes, perhaps we don't always see the full range of emotions. Hopefully, the average cat owner does not see too many negative emotions in their cat. But it's certainly worth paying attention to your cat's body language to learn more about how to tell what they're feeling. That's useful information that will tell you what your cat does and doesn't like. (But don't stare, because that makes cats uncomfortable.)

Throughout this book, you will find a discussion of what feline science can tell us about what cats need to be happy, and every chapter except this one ends with a series of tips to apply the science at home. At the end of the book, you will find a checklist for a happy cat that you can use to help you consider what you're already getting right and where you could make changes to improve your cat's welfare. Please note that none of this information is a substitute for a professional opinion. If you have concerns about your cat, see your veterinarian. If they rule out medical issues, then a veterinary behaviorist or cat behavior counselor may be able to help.

2

GETTING
A KITTEN
OR CAT

....

WHEN HARLEY FIRST came to live with us, we confined him to one room (our bedroom) for a few days while he settled in so that he wouldn't feel overwhelmed. Although it didn't take long before he was happily lounging on the bed, it soon became apparent that he was not a lap cat. I felt sad about this. But I realized that he was trying to be near me in his own way. When I went to sit on the settee, he would come around the edge of the room and settle underneath the settee, just below where I was sitting. Although it was not at first obvious to me that he was there, I thought this was his way of being close to me and feeling safe at the same time. To test out this theory, I moved to the other end of the settee. Sure enough, he moved too so that once again he was just underneath where I was sitting. And if I moved to the chair, he moved too, to be under the chair. He was the opposite of a lap cat: an under-chair cat. Of course, as he settled in more, he no longer felt a need to hide and became an occasional lap cat.

Soon after, when we brought Melina home, we again kept her just in our bedroom to begin with. It took her a little longer to settle in. Every cat will begin to feel at home in their own time. I remembered the very first time I brought an adult cat home to live with me, and how he spent all of his first week hiding. I didn't see him at all during that time, but I could see from his food bowl and litter box that he came out of hiding while I was out of the house or asleep. Giving him space to hide and not trying to force him to interact meant that he could settle in, and ultimately he was actually a pretty bold cat. So we let Melina take her time, and once she settled in, it became clear that she is a bold cat who will happily greet visitors to our home. And if they put their purse on the floor, she will sniff it and try to put her head inside it in case there's anything interesting there.

Bringing home a kitten rather than an adult cat can often feel easier, as they are less likely to be as fearful (although you can get shy kittens); however, you have to do a lot more ahead of time to kittenproof your home. Kittens don't understand that electrical wires aren't for chewing and your sneaker laces aren't for chasing. If you are planning to get a cat, look carefully at your home to spot and solve potentially dangerous situations beforehand.

How to kittenproof your home:

1. **Identify small spaces where kittens aren't safe or from which you can't easily retrieve them.** Kittens can climb and get into very tiny spaces, so you want to make sure that little

holes are blocked off or filled in, such as next to the washing machine, beneath kitchen cabinets, or anywhere they could get trapped inside cupboards.

2. **Place all houseplants and bunches of flowers completely out of reach of kittens.** Check to make sure the plants in your home are safe for cats—and get rid of any that are toxic. Some common plants and flowers are toxic, including lilies, peace lilies, money plants (also known as jade plants), aloe vera, daffodils, and hyacinths (and many more—this is not a comprehensive list). The American Society for the Prevention of Cruelty to Animals (ASPCA) has an online database that you can check.

3. **Tidy away anything that your kittens might break (such as vases) or that might injure your kittens if they try to chew them (such as batteries).** Look at the cords for blinds and lamps and find a way to tidy them up. Ensure they are the kind that will come apart if pressure is applied, as this protects toddlers as well as kittens from potentially terrible accidents. Similarly, some cat toys, such as wand toys or any toys with a string that the kittens might get tangled in, also need to be tidied away except when you are there to supervise play.

4. **Check window screens for holes and secure window openings.** Again, kittens can get through very tiny holes, and many screens are not sturdy enough to keep a kitten (or adult cat) contained. If you have balcony railings or bars, be sure to kittenproof them or prevent access.

5. **Ensure the doors to the washing machine and dryer are kept closed at all times.** Make a habit of checking inside these

machines before using them—especially the dryer, which, since it's often warm, makes the kind of place where a cat might want to hide.

6. Ask everyone in the household to **keep the lid of the toilet down** so that your kitten does not fall in.

7. **Keep your house tidy** so your kitten does not try to eat or play with stray bits of food, elastic bands, paper clips, etc. If swallowed, these items can cause a blockage and require emergency surgery.

8. **If you have a dog in the home, make sure they are initially separated from your kitten or cat.** And give your feline lots of high places and hiding places. See chapter 9 for more on introducing dogs and cats.

CHOOSING KITTENS OR AN ADULT CAT

BREEDERS, FRIENDS, AND neighbors are common sources of kittens. Rescues and shelters can be great sources of both kittens and adult cats. For adult cats with medical conditions (such as diabetes), some organizations will cover the ongoing costs of medication. Whether you're getting a kitten or a cat, make sure they look healthy. Check to see that they have received appropriate vaccinations and parasite treatment for their age (see chapter 6). Check that their ears, eyes, and butt area are clean, that the coat looks clean and healthy, and that a kitten is lively and moving around normally. Kittens from an animal shelter may already be spayed/neutered; otherwise, it's advised to spay/neuter kittens before 6 months of age to prevent unwanted litters.

Always see kittens with their mom and litter, and ask about when they were born, whether they are friendly, where they have been living (it should be in the home, not a garage or outbuilding), and what the breeder has done to help them be well socialized (see the next section for more on socialization). For a pedigree kitten, find out which inherited disorders are common in the breed, and ask the breeder what health checks they've had done on the mother and father. Expect a good breeder to talk to you about these issues, and to ask you lots of questions to ensure you are a good home. Some organizations such as the Royal Society for the Prevention of Cruelty to Animals (RSPCA) in the UK have a kitten checklist that you can use.[1] Don't bring a kitten home until they are at least 8 weeks old.

It's a good idea to get two kittens instead of one, and indeed, some shelters and rescues will insist on this (although they may make an exception if you already have a friendly adult cat at home). Two kittens can keep each other company and be playmates, they will continue to learn from each other, and they can engage in normal feline behaviors with their fellow kitten.[2] As well, they will be more accepting of other cats when they are older.

Adult cats make great pets, as they already know how to behave in the home and are less active than kittens. Find out what you can about the previous life of the cat, as it will be an easier transition for them if your home is somewhat similar to the kind of home they have experienced before. For example, shelters and rescues will often be able to tell you if a cat is used to living with children or with a dog, if they are the kind of cat that is used to and wants outdoor access, or if they are used to being indoors only. If you want more than one cat, shelters often have bonded pairs of cats waiting for a home together.

Although pedigree kittens come from breeders, you can also find pedigree cats in shelter and rescue. A study found pedigree cats are rated as more friendly than moggies (mixed breeds), and there are two main reasons this could be the case.[3] Dr. Lauren Finka, one of the authors of this study, told me, "It might be that because there has been more selection of these types of cats, they're actually further removed from their wild counterparts, so they're less likely to have been the result of a tom (unneutered male) sneakily mating with a female. These are cats that have been controlled and selected in a way that we've done on purpose. So they have probably quite different basic temperaments as a whole compared to your average moggy, which might be second-generation street or feral cat." The other reason, she said, is that the kind of person who gets a pedigree cat is probably different from the kind of person who chooses a moggy, and they may interact with their cat and manage their cat in different ways. There's another possible reason: pedigree kittens typically go home at 12 weeks rather than 8, and that extra time with the mother and siblings may make a difference too.

When choosing a cat or kitten, Finka says, "it's really important to think about why you want a cat and also what kind of environment you will provide the cat with. Those two things will really guide you in terms of knowing what cat is right for you." Most people want a friendly cat, so asking questions about how the kitten is being reared or the background of an adult cat will help you to find a cat that is a good fit for your family. "If you have a busy lifestyle, if there's a lot going on, if there's lots of coming and going—is that something that they could cope with?" says Finka. "Usually younger cats will adapt better to a new environment, so that's also something to think about if you think that your house is a bit chaotic and you really want a cat that is going to cope well with that. Some people, on the other hand, may be very happy with a cat that is quietly resting next to them on the sofa and isn't very vocal, doesn't really want a lot of fuss or attention. Again that's important to recognize because the last thing you want to do is go and adopt a hypersocial, very interactive, very vocal animal; that again wouldn't be a good match."

Some cat breeds, such as the Persian and Exotic Shorthair, have short, squashed faces, which is known as brachycephaly. One study found that cats with flat faces are more likely to have respiratory issues and/or more likely to make noises such as snoring when asleep.[4] Brachycephaly is also associated with eye and dental issues, and 80 percent of show Persians and Exotic Shorthairs were found to have pinched nostrils, while 30 percent had entropion (a condition in which part of the eyelid turns inwards).[5] When scientists analyzed cats' facial expressions, they found that many brachycephalic cat faces look like the cat is in pain even when they are not.[6] The most affected breed was the Scottish Fold. These results suggest that breeding for particular features (a round face and big eyes) has affected their ability to communicate. It's possible that this vulnerable look makes people want to take care of them or be less able to recognize when care is or isn't needed.

Another concern is breeding cats with wildcats. The Bengal cat, originally created by breeding domestic cats with Asian Leopard cats, is now a domestic breed, although some may be difficult to manage in a home environment.[7] Wildcats should be wild, and not be bred with domestic cats.

Some cat breeds have a longer head shape (called dolichocephaly). Research has shown that people who have a cat with a more extreme head shape tend to have "brand loyalty" and prefer other cats with a similar-shaped head.[8] In this study, most people preferred a normal or slightly longer skull. The most preferred cats also had green or blue eyes, a medium/long coat rather than a short one, and were blue/gray, ginger, or tabby.

People often think that coat color is linked to personality, and this idea may influence their choice of cat. But although people think of tortoiseshell cats as more trainable (and also more intolerant), and cats with orange or two-colored coats as friendly, no independent evidence suggests this is the case. In fact, one study of pedigree cats found that personality differences are mainly due to breed, not coat color.[9] Another study found that cat breeds can be grouped into four main groups, that different breeds do have different behaviors, and that some behavioral traits have a heritable component.[10] Several long-haired breeds—the British Shorthair, Persian, Birman, Ragdoll, and Norwegian Forest—were rated as the least aggressive, least fearful, and also least likely to be extroverted. In this case, extroversion meant both sociable to people and very active. Russian Blue and Bengal cats were found to be the most extroverted and also the most fearful, and Turkish Vans were considered the most aggressive. Burmese cats were the least shy, and Korats were most likely to seek contact with people.

"If we promote and teach kitten socialization the way we do with puppies, the world would be a better place for cats. The socialization window—open from 3 to 7 weeks—is so important for them and what they learn about the world. We need to provide kittens with good experiences during this time. Introduce them to different people, environments, objects, animals, and handling. Think of how different their experiences and life would be if we did this. Early socialization will help them not be afraid of new things or changes in their environment, which are bound to happen. It's so simple, yet we don't often think of the importance of cat socialization like we do with dogs. Maybe that's it, we need to change the way we think about cats."

—KIM MONTEITH, manager, Animal Welfare, BC SPCA

THE SENSITIVE PERIOD FOR SOCIALIZATION

THE EARLY WEEKS of a kitten's life are an important time for their brain development and their later behavior as adult cats. The time between 2 and about the end of 7 weeks of age is known as the sensitive period for socialization. It used to be known as a "critical" period, because it was believed that the right experiences during this time were essential for normal development, but now it's called "sensitive" to reflect the fact that changes can still occur in the brain later.

Kittens are typically not homed until after the end of the sensitive period, 12 weeks of age for pedigrees and 8 weeks for shelter kittens. Many animal welfare organizations, such as the RSPCA in the UK, recommend that kittens should not be taken away from their mothers before 8-9 weeks of age because of the risk of behavioral issues if they are taken away younger. It's best for kittens if there is a good match between that first home and the home they will go to permanently, as it means they will get used to normal home life and the sights and sounds that go with it. If instead your kitten was raised in a barn, they won't have been exposed to normal house experiences and may not have met many people. If kittens don't have the right experiences during that early time, their later behavior can be affected, and they may not grow up to be as friendly and sociable as they could have been had they lived in a home and been well socialized. Kittens should be handled by at least four different people during this time, and the handling needs to be a positive experience.[11]

The idea that kittens have a sensitive period was first noticed from research in which kittens were placed with rats and other animals at an early age. Then, in 1970, veterinarian Michael W. Fox identified that the sensitive period in kittens seems to start from about 17 days.[12] We learned a lot about the sensitive period in kittens thanks to work by Dr. Eileen Karsh, who raised kittens in her lab at Temple University in Pennsylvania. There is a lovely account of these studies in Thomas McNamee's book *The Inner Life of Cats: The Science and Secrets of Our Mysterious Feline Companions.*[13] McNamee tracked down the now-retired Dr. Karsh to learn more about her research.

In one study, Dr. Karsh had different groups of kittens who were handled for a four-week period starting from the age of 3, 7, or 14 weeks. The handling consisted of someone putting the kitten on their lap and petting them for fifteen minutes each day. Then, at 14 weeks of age and subsequent intervals up until 1 year of age, she placed them on a person's lap. The kittens who had been handled from ages 3 to 7 weeks stayed for the longest, showing that they were the most comfortable with being near people. She also timed how long it took the kittens to approach a person sitting on the other side of a room. The kittens who had been handled from 3 weeks were the quickest to approach.

One large study looked at the effects of early weaning—taking a kitten away from their mother and sending them to a new home prior to 12 weeks of age.[14] The study asked owners of 5,726 pet cats about their cat's behavior and the age at which they had been weaned. When cats were weaned before 8 weeks old, they were more likely to be aggressive to strangers and more likely to have a behavior problem compared with cats who had been weaned aged 12–13 weeks. In contrast, the cats who were weaned later, at 14–15 weeks, were less likely to be aggressive to strangers and less likely to groom excessively. Cats who were weaned as adults, or who in fact were never weaned (i.e., they continued to live in the same home as their mother), were not only less likely to be aggressive to strangers but also less likely to be aggressive to family members and other cats. Because this was a questionnaire study, it does not prove that early weaning was the cause of behavior issues; however, it was an unusually large study with a large number of cat breeds included (over 40), as well as mixed breeds. Overall, although more research is needed, these results suggest that it is better for kittens to stay with their mother and the rest of their litter until 14 weeks of age, later than is currently typical.

How can you tell a kitten's age if you're not sure? Shelters and rescue centers are used to estimating the age of stray kittens, but perhaps the easiest way is to weigh them in pounds.[15] Up until 5 months, a kitten's weight in pounds is approximately the same as their age in months; for example, a kitten who is 12 weeks old (3 months) typically weighs about 3 pounds (assuming they are not skinny as a result of being starved).

Once kittens have passed the end of the sensitive period for socialization and go to live in their new homes, it is still important to ensure that they have a wide range of positive experiences and that they do not get too stressed. These positive experiences will help them to build on the learning that took place during their socialization and to generalize their knowledge of the world. They will also be able to habituate to things in your home, such as the noises of the dishwasher, microwave, and washing machine, that they don't need to pay attention to.

"Education of people about cat behavior is the biggest key to helping cats around the world. Knowing how cats 'tick' and what they like would be helpful. Teaching this at Kitten Kindy classes is the easiest and most effective way, besides being lots of fun for kittens and owners alike. Cats are social animals but that does not mean all cats are socialized. Being socialized means that the individual accepts the close proximity of others—cats and people. It does not mean they have to like all others—sounds like us really, doesn't it? The socialization period—when it is easiest to help kittens learn about their world—occurs between 3 and 7 weeks of age. So the way the breeder raises the kitten is so important. However, socialization can occur at any age.

So what can make a difference to cats during and after the socialization period? Many people want cats to have a best friend, but many (most) cats are more suited to being an only cat. Some other behavioral needs of cats that are not widely known include: cats prefer their food and water to be a good distance apart. The ideal litter tray is 1.5 times the length of the cat, and very few are made this big. Some also have lids and flaps to keep the smell out for the owner. Many cats will cope but they don't necessarily like them, and those cats may develop toileting problems as a result. So knowing about the cat's behavioral needs and respecting them would really make the world a better place for cats, and starting during the socialization period is best."

—DR. KERSTI SEKSEL, BVSc (Hons), MRCVS MA (Hons), FANZCVS, DACVB, DECAWBM, FAVA, veterinary behaviour specialist at Sydney Animal Behaviour Service

SETTLING YOUR CAT IN

WHEN YOU FIRST bring home a cat or kitten, set up one room for them, such as a bedroom or bathroom, with everything they need. This means they can settle into that room before you let them explore the entire living space, which might be too stressful for them. I asked Dr. Karen van Haaften, veterinary behaviorist at the BC SPCA, for her advice on bringing home a cat who is fearful. She said, "I'd be prepared to have them hunker down in there for at least a day, possibly three or four days depending on how quickly they adjust to the environment. But until they're eating, showing social behavior with people, showing lots of relaxed behavior, getting lots of good-quality sleep—until they're at that stage, I don't think I would recommend giving them access to a larger space, because that larger space could actually be very overwhelming for them. And the people's interactions are important too. The whole family, of course, is going to be really excited that there's a new kitty coming home, but we don't want it to be that the cat comes home in the carrier and everybody crowds around the carrier and tries to pet the cat and be the first to pick them up and kiss them. Cats are going to be overwhelmed by that. The way that I usually recommend people bond with a new cat is just to spend time in the same room with them, maybe reading a book or watching a show on a tablet or something, not really invading the cat's space but letting the cat come to them."

She says that a cat who has come from a hoarding situation will need a lot more time to settle in. She describes the situation: "We'll bring in 50-100 cats from somebody's home and of course if those cats have just been reproducing freely in the house they can't all have good social contact with people." Because these cats have not had those early socialization experiences, the shelter staff will do a lot of work with them to prepare them to go to a new home. These cats will need more time to adjust to that one safe room when they first arrive in your home. The shelter will also offer new owners some guidance on how to help the cats settle in. "We might have to teach them some of our DSCC [desensitization and counter-conditioning] to petting protocols from the shelter to help remind the cats that, oh yes, petting is often and in this room with these people too, just like it was in the shelter," says Dr. van Haaften. She also said that the cats may not know how to use a litter box but typically learn quite quickly, and they may always need sedation to cope with a vet visit, because the experience is so far outside of their comfort zone and they have not experienced it before. But while it's important to know about these possible challenges, they can still make lovely pets in the right home.

APPLY THE SCIENCE AT HOME

- Don't bring a kitten home before 8 weeks of age. Always ask to see them with mom and the rest of the litter, check that they have been living in a home environment (not a garage or outbuilding), find out about their early socialization, and check that they look healthy. Get two kittens instead of just one.

- Even though the sensitive period for socialization is over when you bring your kittens home, it is important to continue to give them positive experiences to build on what they already know. As well, prevent negative experiences, including by kittenproofing your home.

- If adopting a rescue cat, find out what's known about their previous life and consider how it will help the cat fit into your family home.

- Prepare just one room for the cat to arrive to, with everything they need. Wait until they are comfortable in this room before giving them access to other parts of your home. Be patient; moving to a new home is a big deal for the cat.

3

HOW TO SET UP YOUR HOME FOR A CAT

·················

O N HER FIRST day in our house, Melina found the perfect hiding spot, one that was extremely annoying. We knew she was hiding somewhere, but it took some time to fig-ure it out. I looked under the bed, in the closet, beside the chest of drawers, next to the wicker chair my grandad made for me when I was a kid, under the duvet, and under the bed again. Then, looking under the bed for about the tenth time, I noticed a little piece of fabric dangling down in the corner where the sheet was no longer tucked in. Could she be inside the box-spring mattress? I kept very still and very quiet, and eventually I heard a little rustle of movement. The answer was yes.

Somehow Melina had found a tiny hole and made it bigger, and into the mattress she went. And stayed. This was especially annoying at night-time, because it was when we were trying to sleep that she felt safe enough to come out. Having a little cat wiggle her way around the box springs inside the mattress from under your pillow to the place by your toes where she found her way in is not conducive to sleep at all! But she was too scared to come out during the day, so we had to put up with it until she settled in. I'm a light sleeper and so those first few nights, I did not sleep well. Gradually, Melina began to hide in the mattress less and less, and after about three weeks, when she was rarely going in there, it seemed all right to finally close up the hole so that she no longer had access. By now, she had other places she was happy to hide, and she was also wandering around as if she owned the place.

It's understandable that a new cat or kitten might be nervous in your home. That's why it's best to keep them confined to one room for the first few days. But it's important to set your house up right for your cat(s) throughout their lives, not just at the beginning. Things that seem perfectly fine to us might be all wrong from the cat's perspective and, unfortunately, can make your cat stressed and cause behavioral issues. As discussed in chapter 1, a survey of experts on all things feline, published in *Veterinary Record*, found the single biggest welfare issue affecting pet cats is social behavior problems caused by a poor environment at home.[1] Stress and anxiety can contribute to a range of behavior problems, including house soiling and aggression.

The good news is that following five basic rules can help ensure that cats will be happy with their home. These five pillars of a healthy feline environment were described by the International Society of Feline Medicine and the American Association of Feline Practitioners.[2] Together, they contribute to the domains of the physical environment and to behavioral interactions with the environment and with people. Let's go through them in turn.

1. A SAFE SPACE

WE OFTEN THINK that if we weren't around, cats would catch mice and so on for food. We don't tend to think about cats being prey too, but they are. This is most obvious for those in North America, where coyotes and cougars may predate upon cats. Cats are concerned about their welfare, and they don't like confrontation: their coping mechanism is to hide.

Dr. Wailani Sung, veterinary behaviorist at the San Francisco SPCA and coauthor of *From Fearful to Fear Free*, explains that the fact cats are prey affects their behavior: "Cats are predators to small prey but they're prey to larger predators. And that's why some cats are always, like, twitchy or spooky and easily startle." This is why it is essential for us to provide them with safe spaces where they can hide or perch away from anything stressful, or simply relax.

Just how important safe space is was shown by a study published in *Behavioural Processes*, in which shelter cats were put in a choice chamber with a space in the middle and cat flaps leading off to different compartments.[3] In one compartment was a hiding box; in another, a perch; in another, a toy; and the final compartment was left empty as a control. The cat's food and their litter box were in the middle space. The scientists monitored where cats chose to spend their time over seven days in the chamber. Although every compartment was accessed roughly an equal number of times, cats spent the most time in the compartment with the hiding box. Some cats spent more time in the perch than others. The main conclusion was that a hiding place is not just nice to have; it seems to be a basic need for cats.

Harley enjoys spending time at the top of his cat tree. ZAZIE TODD

Cats like opportunities to explore and be high up. JEAN BALLARD

A good hiding place is somewhere that is just the right size for a cat, so it is nice and cozy (not spacious!). Because cats like to perch and survey their surroundings, high-up places—such as enclosed spaces or perching spaces at the top of many cat trees—can be good hiding spots. As most people are aware, a cardboard box can be a great safe space for a cat. Try turning the box upside down and cutting a cat-sized hole to make an access point. If you look around your home, you will probably identify places where your cat already likes to hide or nap, and other spaces that you can easily transform. Add a blanket or cat bed to the space under beds or settees, or to part of a bookshelf or a shelf affixed to the wall (ensure it is stable and will not fall). Maybe somewhere in your linen closet or at the back of your wardrobe there's a nice little cat-sized space, or a pile of linen that your cat can burrow into. A sunny spot by a secure window is a great place for a perch. Take a look at each room in your home and see if there is a suitable safe space for your cat in that room. If the room allows, make several.

"I think the single most important improvement in the care and welfare of cats is, quite simply, a better understanding of a cat's environmental necessities. This greater grasp needs to be industry-wide. There is a distinct disconnect between what cats need and what is being offered to feline caregivers to help meet those needs. A clearer understanding of a cat's physical and emotional needs must be improved across the board from veterinary professionals to pet product designers and cat guardians alike. Forcing cats to comply with our human world and expectations sets them up to fail. Less environmental stress leads to less physical illness and fewer behavior problems. There is a recent trend to change the term 'environmental enrichment' to a more fitting 'environmental needs,' and I embrace that trend. Many 'behavior problems' would never become problems at all if cats were simply provided with an environment that embraces their inner 'catness.' They need bigger litter boxes (and more of them). They need taller and more abrasive scratching posts, not these ridiculous carpet scraps that hang from doorknobs. Cats need to be challenged and offered a stimulating world that evokes their inner predator while simultaneously providing the safety and security they crave. The movement is growing as the public demand for more knowledge and better products increases, but we still have a long way to go towards making our homes more feline friendly."

—INGRID JOHNSON, certified cat behavior consultant at **Fundamentally Feline**

2. MULTIPLE AND SEPARATED RESOURCES

THE NEXT PILLAR of a healthy feline environment is that your cat needs multiples of each resource, and they should be separated from each other. Take a moment to think about what might be a resource to your cat. Does your list include their litter boxes, food, water, scratching post, bed, and perhaps some toys? These are the kind of thing we mean by resources. When multiples are available, it means that if your cat does not feel safe accessing the resource in one location, they have another one to try. Resources should also be separated from each other, because if two things are right next to each other, in the cat's eyes they might as well be one. One common mistake people make is to put two litter boxes adjacent to each other in a room. If the cat doesn't feel comfortable accessing one of those litter boxes—perhaps because it's in a bathroom and someone is taking a shower, or another cat is there—they are not going to feel any safer using the other one in the same room.

In a household with more than one cat, one cat will sometimes block the other cat's access to a particular resource. For example, you may see the cat lying in the doorway of the bathroom where the litter boxes are. This makes it very difficult for the other cat to go into that room. So it is especially important to have separated resources in a home with more than one cat. A common rule of thumb is to have one litter box per cat plus one extra. Sometimes cats will form social groups (see chapter 9). Cats within a social group will often lie touching or very close together. These cats will likely be willing to share resources. But if more than one such group exists within a multi-cat home, these different social groups should not have to share.

If your cat has outdoor access, try to ensure that they have access to their resources outside too; for example, by providing a water bowl. It may not be possible to provide food outdoors, as it might attract rats, raccoons, or other wildlife. But if your cat has access to your yard or balcony, you can certainly ensure that your cat has safe spaces there, such as under a patio table or behind plant pots. Also make sure that their point of entry to inside is a safe space with clear lines of sight.

3. OPPORTUNITIES FOR PLAY AND PREDATORY BEHAVIOR

ANOTHER PILLAR OF a healthy feline environment is that the cat should have opportunities to play and show predatory behaviors (those that would be involved in catching a mouse, for example), which will help satisfy their hunting instinct. Play can be social (with you or other cats) and also nonsocial, as cats will happily play on their own with toys, such as toy mice or balls. Since play involves predatory behaviors, cats like toys that seem like prey, such as those with feathers or tails. As well, if you are playing with your cat (e.g., with a wand toy), try to move it as if it were prey. For example, prey might freeze and then run away, but it isn't likely to go right up to your cat. Don't feel awkward if you're having trouble engaging your cat; just experiment with different ways of moving the toy. It's not a good idea to let kittens play with your hands, as this activity won't be so much fun for you when they become an adult cat with stronger, sharper claws. Redirect them to toys instead. Always tidy wand toys away when you're not around, as they can be a danger to cats, and don't leave objects (like rubber bands) lying around that cats might want to play with but that are dangerous to them.

Cats are notoriously fussy, but it's a good idea to provide a range of toys. "I recommend representing all of the major prey groups in your toy chest," says Dr. Mikel Delgado, certified applied animal behaviorist and cat behavior consultant at Feline Minds and staff scientist at Good Dog. "Birds, mice, snake-like or lizard-like toys, bug-like toys, and probably a larger mammal like a squirrel-like or rabbit-like toy. And I'm always emphasizing the importance of what we call interactive toys—the teasers, the stick with the string with something dangling on the end." Some cats prefer to specialize in certain kinds of prey (or toys) whereas others are generalists and love them all.

Cats will get used to toys, as a study that tested cats playing with objects showed, but they may start playing again if the toy changes.[4] The hypothesis is that with prey items, cats get bored unless the prey changes or another prey item comes along. Dr. Delgado, who was not involved in this research, explains: "Cats continue hunting when they are getting feedback from the prey that it's going well. That includes things like the prey's body changing form, like feathers flying off or skin getting ripped; basically destruction of the body of the prey animal. When we play with toys, we think we don't want to keep buying toys so we buy these very sturdy toys that won't fall apart. But cats might actually like the sensory part of the toy being destroyed." This is probably why cats enjoy shredding tissue paper, newspaper, and cardboard.

Ingrid Johnson, certified cat behavior consultant at Fundamentally Feline in Atlanta, Georgia, agrees that toys should change often. "What is the motivation to chase a feather toy that has been lying around on the floor for weeks? Dead, done, boring!" she says. Keep toys in a box or toy chest and bring out a few at a time. Or store them with a scent that your cat likes (catnip, for example) to make them seem new. Johnson also stresses the importance of small environmental changes, such as adding a cardboard box, a brown paper bag, a box of leaves from outside, or a pile of tissue paper on the floor.

The average cat has seven toys, according to a study published in *Journal of Veterinary Behavior*.[5] The most common is a furry mouse, owned by 64 percent of cats. The other most common toys were catnip toys, balls with bells, stuffed toys, a scratching post, boxes, and balls without bells. Most of these toys provide opportunities for play and hunting behaviors. Very few of the cats in this study had food-puzzle toys (more on these in chapter 10), but they should also be part of your cat's repertoire. Although most owners (78%) had these toys available all the time, remember to change the toys regularly so your cat doesn't get bored. Some owners in the study played with their cat only once a month, but 64 percent reported playing with their cat twice a day, and 17 percent reported daily play sessions. Typical play sessions were five minutes (33% of owners) or ten minutes (25%). Most cats would probably prefer longer play times.

Veterinary behaviorist Dr. Beth Strickler of Veterinary Behavior Solutions is one of the authors of this study. She told me that every cat is an individual and will have their own preference for type and duration of play. In general younger cats will need more play, but some older cats will also be very playful even if they need slower-paced sessions. Dr. Strickler says, "It's less about the amount of time and more about finding out what the cat likes to engage in with the person, then providing the opportunities for the cat to engage when they would like to." Suitable toys for solitary play are also important.

The cats in this study were selected because they went to the vet for a reason other than behavior; in fact, however, 61 percent of the cats were said to have one of six common behavior problems. And only 54 percent of owners had mentioned the problem to their vet. The two most common behavior problems were aggression towards the owner (36%) and house soiling (24%). It is especially concerning if house soiling is not mentioned to the vet, as there are potential medical causes and many options for treatment and management (see chapter 11). The number of toys the cat had and how often the owner played with them were not related to behavior problems. However, owners who played with their cat for at least five minutes at a time reported fewer problems compared with those who played for only one minute.

One way to determine if your cat needs more play is to observe what happens when you stop. At the end of the play session, put the wand toy away and give the cat a toy mouse, food toy, or treat, as if they "caught" something. Ideally they will have a snack, do some grooming, and be ready to take a nap. If they seem frustrated or are still trying to chase your ankles, it's a sign they need more play in their life.

It's important for cats to have opportunities to play with toys on their own, like Clancy with this catnip toy, and to play with their person.
JEAN BALLARD

4. POSITIVE, CONSISTENT, AND PREDICTABLE INTERACTIONS WITH THEIR PEOPLE

CATS NEED POSITIVE and predictable social interactions with the people they live with, but always give the cat a choice. A common conflict is that the person would like to have one long petting session, whereas the cat would typically prefer multiple short petting sessions throughout the day. This preference can vary with age. Kittens and young cats tend to prefer longer interactions, and as they get older and reach social maturity (around 2–3 years), they may prefer shorter sessions. As well, if cats develop mobility issues such as arthritis as they get older, this may change their interactions. They may prefer to sit next to you on the couch, or sit on the floor and look hopeful that you will pick them up so they can sit next to you without jumping. But some cats may prefer not to be picked up even if they used to like it. Never assume changes are simply due to age. Always check with your vet in case of an underlying medical condition.

If you have more than one cat, ensure that each cat gets some one-on-one affection from you. You might notice that each cat has their own individual preferences; for example, one may prefer a slightly longer petting session than the other, or one cat may easily bite if petting goes on for too long. Pay attention to each cat and try to learn to provide what they like. If someone in your household is suddenly out a lot more—because of a new job, for instance, or because they traveled—expect your cat to notice. They may want more fuss from the family members who are still at home, or conversely may be pining for the person they aren't seeing as much and be demanding of them when they get home. Don't neglect your cat's needs for social interaction. For more on how to pet a cat, including where they prefer to be petted, see chapter 8.

"One thing that would make the world a better place for cats is if cat owners understood that, although domesticated, companion cats retain the instinct and desire to perform normal feline behaviors. These behaviors include scratching and scent-marking; seeking, hunting and stalking; and maintaining a secure territory. Until fairly recently, pet cats were free to roam their neighborhoods at will. This freedom allowed them to maintain a larger territory, to hunt and stalk prey, climb, scratch, problem solve and keep physically fit—in essence, to be a cat.

Pet cats, especially those confined to the home, must have appropriate outlets for the expression of these normal feline behaviors—without them, problem behavior, such as destructive scratching of furniture and aggression towards people/other animals in the home, is common. Stress and anxiety, especially common in multi-cat homes where cats compete for access to valued resources, can lead to inappropriate toileting and territorial marking—a common reason for relinquishment to animal shelters. Providing cats with appropriate outlets for normal feline behaviors should include: the provision of scratch posts and cat scratchers; cat towers or high shelving to provide a safe place to retreat to; ready access to several litter trays (especially important in multi-cat households); and safe access to outdoor space if possible. The opportunity for daily play with toys that mimic prey provides an outlet for hunting and stalking behavior, which can reduce aggression towards people or other animals in the home."

—KATE MORNEMENT, PhD, applied animal behaviourist and consultant at Pets Behaving Badly

5. AN ENVIRONMENT THAT REFLECTS THE CAT'S SENSE OF SMELL

MOST PEOPLE DON'T realize just how amazing cats' noses are. It's dogs who typically get the credit for being brilliant at sniffing. But cats have great noses, along with something called a vomeronasal organ (VNO).[6] So it is safe to say that scent and pheromones are very important to cats. A pheromone is a chemical signal that can be detected by the VNO, which is located above the hard palate of the mouth. You've probably seen your own cat doing the Flehmen response: it's when the cat has their mouth open and their lips may be pulled back in a kind of sneer. They are investigating something with their VNO. Because air does not reach the VNO, scent molecules must be absorbed into mucus and can get there via two openings in the roof of the mouth, just behind the incisor teeth.

We don't fully understand all about pheromones in cats, but what we do know so far is fascinating. Cats secrete pheromones on the head between their eyes and ears, on the side of their forehead, in their cheeks, under their chin, at the corners of their mouth, between the pads on their paws, at the base of their tail, in their anogenital area, and around a female cat's nipples. These different pheromones have different purposes.

Pheromones and scent are important from the moment a kitten is born. The mother cat secretes pheromones from glands around her nipples. From 1–2 days old until 32 days, kittens within a litter have a preference for a particular nipple while nursing (something called "teat constancy"). It's not known exactly how they know which one is "their" nipple, but it's possible that the mother's pheromones plus their own scent left from previous suckling helps them find it. The nest has its own scent, which is a mixture of secretions from the mother along with hair, urine, and saliva from the kittens. Kittens are born with their eyes and ears closed but can find the nest by following this scent, along with their sense of touch and the feelings of warmth. The nest odor is thought to reduce stress, have a calming influence, and improve the kittens' well-being.

If you have more than one cat, you've probably seen them sniff each other. Scent plays an important role in feline social behavior. Cats who are within the same social group will rub their head on each other or even sometimes their whole body (called allo-rubbing), and this is believed to contribute to building up a group scent. Pheromones are also in poop, and a study of just one cat discovered that bacteria in the cat's anal glands contribute to that particular scent.[7] Cats can identify their own poop from that of other cats, and can also tell the difference between poop from familiar and unfamiliar cats, as shown by how long they spend sniffing the poop.[8]

When cats poop in their own yard, they typically bury it. In other locations they sometimes leave their feces uncovered, which is called middening. Although we don't fully understand the reasons for leaving feces uncovered, one possibility is that they are leaving a visual and chemical signal that this is "their" territory, in order to keep other cats away.

When cats rub their head on things, they deposit pheromones from the glands in their cheeks. This is called bunting. Facial rubbing can leave five different pheromones, known as F1, F2, F3, F4, and F5. We don't know what the F1 and F5 pheromones do. F2 is thought to be connected to sexual behavior—for example, when a tomcat rubs his head on a receptive female cat, he deposits F2 pheromone. F3 is considered to be territorial in nature. We know most about F4, which is thought to maintain group cohesion via allo-rubbing amongst cats who are friendly. As well as allo-rubbing, cats in the same social group also rub the same areas in their environment (such as walls and tables in your house), which may help to create a group scent. Similarly, when your cat facial rubs on you, they are depositing pheromones as if you are part of their social group.

Because of scent's role in social behavior, you can use it to build trust and create a feeling of safety. For example, when introducing two cats, you can introduce one's scent to the other long before they actually meet. But remember to give cats a choice. Do not force a cat to interact with the scent of another cat (for example, don't rub it on them), as they may not like it. Similarly, because cats will put their own scent on their bedding, you should not wash all of the cat's bedding at once. Leave some of the bedding that still smells of them, so they don't feel stressed. Scent also explains why your cat may sometimes choose to relax on dirty laundry you've left on the floor, like when Melina lies on a sweater my husband neatly folded for her.

Since some of these scents make cats feel safe and secure, we need to preserve them for the sake of our cats. Avoid using strongly scented cleaners near places where the cat likes to relax, because they may remove the cat's scent and replace it with something the cat finds unpleasant. This would include places like the litter tray, where it's important to use unscented cleaning products and cat litter. You wouldn't want your cat to toilet somewhere else just because you've made the area smell of some artificial aroma that you like but the cat doesn't. Similarly, you might notice that the furniture or wall is marked and paint may even be rubbed off in places where your cat rubs frequently. There is a little mark on the corner of a wall in our hallway where Melina in particular has rubbed her head many times. Because your cat gains comfort from these areas smelling like them, if you can hold off cleaning or repainting as long as you can, your cat would prefer it.

Cats deposit scent from glands in their paws when they scratch. Over time, this scent may build up on the scratched object and it becomes an "olfactory reference point" for the cat. For this reason, providing the right kinds of places for cats to scratch and leave their scent is important for the cat's welfare, as well as to protect your furniture. Reward the cat with a little treat or some petting for using their scratching post. Don't punish them for scratching other places (see chapter 5).

One study published in *Journal of Feline Medicine and Surgery* found that while some cats (especially unneutered males) will often scratch in the house, others never scratch in places their owner might consider inappropriate (especially neutered males and unspayed females with outdoor access).[9] But this study showed that cats will use scratching posts if they are available. Another study published in the same journal looked at the kinds of scratching posts people make available for their cats and found that many of them are lacking.[10] For instance, many posts are not sturdy enough, may be attached to something like a door that moves, and/or are not tall enough for the cat to get a good stretch while they scratch. Scratching posts that hang from or are fixed to the wall seemed to be particularly associated with scratching issues, which suggests many cats don't like this kind of post.

We can learn what cats like from the posts provided in homes where very few scratching issues occurred. Cats seem to prefer sisal (rope) posts, and cat trees with one or more levels were also associated with low levels of inappropriate scratching. As well, the study looked at how people respond to the cat's scratching behavior. Telling the cat off, removing them from what they were scratching, or redirecting the cat to something else all had no effect. The scratching issues continued. But rewarding the cat with a treat, petting, or praise for using their scratching post was associated with low levels of scratching issues.

Ultimately, teaching cats to scratch in appropriate places comes down to two things: providing scratching posts that your cat likes to use and rewarding the cat for using them. It helps if cats can have more than one post, and since all cats are individuals, some may prefer wooden posts or carpet (especially senior cats). Cats also like to have horizontal surfaces to scratch; there are some great cardboard ones on the market. Another study, again in *Journal of Feline Medicine and Surgery*, gave kittens choices between different types of scratching post.[11] They preferred an S-shaped cardboard scratcher over the other options they were given (which included a sisal post and various other cardboard posts), and it made no difference whether or not it had the scent of catnip or another cat's hair on it.

If you take care to follow the five pillars of a healthy feline environment, you are already well on the way to helping your cat be happy. In the next chapter, we'll look at some other aspects of your cat's environment.

APPLY THE SCIENCE AT HOME

- Make sure your cat has safe spaces in which to hide that are just the right size for them. When you see your cat using such places, don't disturb them—let them hide if they want to, as hiding is a normal behavior for cats and their way of dealing with stress.

- Ensure the cat has multiples of each of their resources, and that they are in separate locations. That way, if the cat doesn't feel comfortable with one location, they have another one they can access instead. This is especially important in homes with more than one cat or with other pets too (such as a dog).

- Ensure your cat has toys to play with on their own, such as toy mice, balls to chase, toys that are scented, a toy that lets them kick out with their back legs, and so on.

- Make time for interactive play with your cat each day. A wand toy is perfect for this. If you have trouble engaging your cat in play, try moving the toy in a different way until they are interested—remember that you are trying to mimic the movement of a prey animal that would run away from the cat.

- Give your cat a choice of whether or not to be petted or to sit on your lap. Every cat is different, but most will prefer to be petted around the head and not at all on the tummy. Keep petting sessions short but frequent.

- If your cat is shy, accept it. Some cats love to be right there with their person, but if your cat prefers to be in the same room rather than on your lap, that's okay too. They are still showing affection in their own way, even if it's more reserved.

- Remember that when your cat rubs their head on things, they are depositing pheromones that help them to feel safe and secure. Try to preserve at least some of them when you are cleaning.

- For cats with impaired vision, scent might be a useful refer-ence point in the environment that helps them find their way around.

4

KEY ASPECTS OF CARING FOR A CAT

· · · · · · · · · · · · · · · · ·

A LOT OF THE wildlife where I now live in Canada is different from where I grew up in northern England. Maple Ridge, like some other parts of Metro Vancouver, nestles up against the deep forests and mountains that make British Columbia so beautiful. Creeks cross the land like ribbons of green, and new streams bubble up in fall or winter when the rains are heavy and the cloud is low. These corridors give wildlife easy opportunities to come down from the mountains and find their way into suburban areas, where a smorgasbord of food waste can be found every garbage day. Since moving here, I've seen bobcats, black bears, cougars, and more coyotes than I ever imagined. When we got a cat, the neighbors warned me that this is not a safe place for cats to be outside. If you'd told me years ago, back in England, that one day I would have indoors-only cats, I would have thought that was cruel and impossible, but now it seems only safe. And even as indoor cats, Harley and Melina still have infrequent wild-life encounters through the window.

One day a couple of years ago, I had just taken some laundry into the bedroom when I glanced out of the window and saw a black bear making his way along the side of the garden. Melina was with me and she saw him too. We watched as he moved along the edge of the trees. When he went out of sight, I hurried to the living room for a better view and Melina came too. The bear ambled into our neighbor's yard, sniffed around their car, and then came back through the trees and started walking towards the front of our house. Melina was on the window ledge, ears flat, tail swishing madly, and she let out the loudest growl you can imagine. It was incredible that such a loud noise came from a tiny cat, but the bear, outside, either did not hear or was not both-ered. He moseyed along slowly, close to the window, and paused to sniff the air with his nose up, moving his head as if to get a bet-ter scent. It was only when our dog Bodger, who was in another room, noticed him and barked that the bear decided to speed up, just a little, and walk along the front of our house and out to the road. All this time, Melina was growling, growling, while Harley, who by now had noticed something was happening, did not really seem bothered.

Melina was very stressed but I wasn't sure how to help her. She was kind of vibrating and it didn't seem safe to try and touch her, so I left her be. I checked on her an hour later, and she had moved back by the bedroom door but was still fizzing and only glanced when I said her name. An hour after that, I was able to call her to me and engage in some petting to help calm her down. There weren't any more wildlife encounters like that—at least, not that we know of—until last week.

One evening, Melina was very out of sorts, stressed, and listening for something. After checking that there hadn't been a distant earthquake, I decided that she must have seen some wildlife again. This guess was confirmed when she would not go downstairs. On these hot summer nights, she had been spending the evenings hanging out on a cat tree by the window in the cool of the basement. She went partway down the stairs and then stopped, listening, looking worried, as if something might be down there. The next morning my husband happened to be doing yard work and came across a large pile of poop right by the window where Melina likes to spend her evenings. It was as if a certain creature came and, unable to get at Melina because of the window, left a strong statement of anger behind. Given the size of that statement, my guess is that it was a cougar who came right up to the door. No wonder Melina was so put out.

For me, the presence of wildlife around our house confirms my decision to keep our cats indoors, although having indoor cats brings its own challenges for good welfare. There is a big cultural divide on this issue between the UK, Europe, and New Zealand (where cats are typically free to go outside for at least some of the day) and Canada and the United States (where indoors-only cats are more common). For some people, this decision is solely about what's right for the cat, whereas for others it is also about perceptions of the potential effects on birds and wildlife. But even this is not as simple as newspaper headlines might make you believe.

INDOORS ONLY OR INDOORS-OUTDOORS?

YOU WILL NO doubt be reading with your existing practices for your own cat in mind and with your own ideas about the wider picture. The decision regarding whether your cat should be indoors only or have access to some daytime outdoors or to an enclosed catio-type area is a very individual one. Veterinarian Dr. Naïma Kasbaoui of the Research Institute in Semiochemistry and Applied Ethology in Apt, France, says: "I really think that they [cat guardians] need to think about the environment and the cat. The nature of the cat: Was this a cat that was adopted as a kitten and was indoors and they just want to give them more space? Or was it a cat that is a stray cat or that has been going out for a long time and so may feel a bit frustrated if the cat was confined indoors after? The second thing is that there is no right answer because it depends on every individual situation."

Although the indoors-outdoors debate is a thorny topic, it's also an area where scientists have done some really interesting research. An analysis of the research on how far pet cats and farm cats roam (known as their home range), published in *Biological Conservation*, found that male cats roam farther than female cats, and whether they are spayed/neutered has no effect on this range.[1] In rural areas where houses are farther apart, cats have a larger home range than in urban areas where houses are close together. In fact, the average home range in a rural area was 14.4 times that in an urban area. This could be because of people, dogs, and cars in the urban area, but it's most likely because more domestic cats there means that each cat has less space for their own home range. Adult cats aged 2–8 years have a larger home range than cats over 8, who tend to stay closer to home, perhaps because it is harder for them to defend their territory. But being socialized to people, and getting regular meals and veterinary care (which happens less for farm cats), did not affect the size of the cat's home range.

One study of pet cats in Denmark, reported in *Animal Welfare*, found that cats who are indoors only or indoors with access to an enclosed garden are more likely to have behavior problems (house-soiling issues, boredom, and damaging furniture) than cats who live mainly indoors but have freedom to roam outside at least part of the day.[2] A Brazilian study also found that cats confined indoors are more likely to have behavior problems and be obese, but have a closer relationship with their owner.[3] This shows the importance of ensuring a good environment if deciding to confine a cat indoors.

Access to the outdoors allows cats to more easily behave in ways that are normal for them, such as stalking, chasing, and doing other predatory activities, to get more exercise, and to spend more time in an environment that is arguably closer to what they are adapted for than indoors in a home.[4] Cats who are indoors only are more likely to be overweight or obese (see chapter 10).

"It would be great for owners and those caring for cats to have an appreciation of the evolutionary history of the domestic cat, and how important this is for a better understanding of this species and their biological and psychological needs today. Many of these needs are still similar to those of their closest ancestors, the North African/Arabian wildcat; a self-reliant predator which is territorial spends a large proportion of each day exploring and hunting, and values solitude and the ability to escape from threats by hiding or getting up high. However, we now expect the domestic cat to live in a world very different from that of its relatives—often restricting their ability to explore and hunt within a complex environment (i.e., by confining them indoors), expecting them to live a much more social lifestyle (i.e., with other cats and with us), and to tolerate a lot of physical handling (i.e., we love to cuddle and fuss over cats). Whilst many cats are able to cope well and live up to our expectations, many may also struggle, either due to a lack of suitable socialization during their early development (i.e., 2–7 weeks) and beyond, other aspects of their temperament, or simply a lack of opportunity to behave as they are biologically motivated to. How we can help is by choosing cats that we think will be able to enjoy the type of lifestyle we have, supplying them with many opportunities for positive cognitive and sensory stimulation, the ability to escape from things they find stressful, and also being careful about how much 'social pressure' we are exerting upon them; providing them with many opportunities to have time alone undisturbed."

—**LAUREN FINKA**, PhD, feline scientist at Nottingham Trent University

CATS AND THE RISKS OF OUTDOORS

CATS WHO GO outdoors face a number of risks. These risks include consuming poisonous plants such as lilies or ingesting toxic chemicals such as antifreeze by drinking from contaminated puddles.[5] It is not known how many cats die of antifreeze poisoning each year, but even a small amount can cause kidney damage and be fatal. If you see signs—vomiting, staggering, or appearing drunk, being sleepy or depressed, finding it difficult to breathe, or having seizures—rush your cat to the vet. Salt used to treat roads and sidewalks in the winter is also toxic to cats (although you can purchase pet-safe de-icers). The relative risk of all these things is not known. Another risk to cats who catch mice and other animals is secondary poisoning if people in the neighborhood use poison to kill rats, mice, moles, or rabbits.

One of the main risks to cats with outdoor access is getting hit by a car. Dr. Naïma Kasbaoui told me: "During my PhD we did a survey in the community. It was a rural community and a town community in Lincoln [UK], a small city. And they all said that the first risk for a cat was road traffic accidents, for a cat that is free to roam." Indeed, a 2004 study involving six veterinary practices in the UK looked at 127 cats who were hit by a motor vehicle; ninety-three survived.[6] Some of the cats who survived had quite severe injuries, in some cases needing a limb to be amputated, and a few needed more than five months to recover. The average time it took for cats to recover physically was forty-seven days, with more severe injuries requiring more expensive veterinary treatment. The owners of thirty-four of the surviving cats (just over a third) said they'd observed a behavior change, such as being nervous about going outside or being more fearful of cars. Around a fifth of the cats' owners reduced their cat's access to the outdoors, and some other owners made no changes but worried more. The study found that even when cats survive a road traffic accident, there can be long-term effects on their welfare.

Another reason some people keep their cats indoors is to protect the cat from wildlife in the area and/or the wildlife from the cat. In North America, coyotes, cougars, owls, eagles, and other wildlife may predate upon pet cats. Dogs roaming free in the neighborhood may also pose a risk. When scientists used radio collars to track free-roaming (not pet) cats in the southwestern suburbs of Chicago, the cats tended to stick to urban areas and avoid the surrounding woods and other natural landscapes where the coyotes live.[7] The study suggests the cats were staying close to houses to avoid the coyotes, although there could have been other reasons too.

Scat provides clear evidence that coyotes will eat cats, even if they aren't the main part of their diet. In Los Angeles, interns at the National Park Service and citizen volunteers collected coyote scat for a study of what coyotes eat, and the scientists also collected whiskers from live-caught or found-deceased coyotes.[8] Since coyotes live throughout the Los Angeles basin, samples were collected from the built-up downtown, from suburban areas where the houses have front and back yards but the hilltops remain undeveloped, and (of whiskers but not scat) from the rural Santa Clarita River Valley where farmland and undeveloped areas nestle. Overall, this area covers a wide range of habitats, including coastal sagebrush, city streets, oak woodland, farmland, and suburban backyards. The analysis showed that in urban areas, 19.8 percent of scats included cats (of course there is no way to determine how many were pets and how many were free-roaming cats). In a much smaller number of scat samples were pet food (less than 3%), domestic dogs, and chickens (each less than 1.5%). In the suburban areas that were studied, cats were in 3.9 percent of scats. In both locations, ornamental fruits and seeds made up a large component of the scat (a surprise to me since I didn't know coyotes ate such things as figs, grapes, and palm fruit). It's possible that the prevalence of fruits in gardens encourages coyotes to the area, putting cats at risk.

Another study analyzed the scat of coyotes in the Denver metropolitan area.[9] The scat was identified visually as looking different from dog poop, so it's possible that if coyotes had been eating similar foods to dogs, the coyote scat was not collected. Of all the mammal hairs found in the scat, pets accounted for only 3 percent, suggesting that coyotes do not eat a lot of dogs and cats. However, from a pet lover's perspective, one is too many. Pet hair was more common in the scat in high-density areas compared with rural areas. Although conflicts between coyotes and pets are more often reported in December and January, no seasonal pattern of pet hair showed up in scat. Hence, the researchers suggest that coyotes kill pets not only for food, but also because pets are a threat or competition in the area (i.e., they kill more pets than they eat).

Coyotes, like cats, eat a lot of rodents, so they may see cats as competing with them for food. However, there was an increase in the amount of pets in coyote scat in March, which is when coyote pups are born. In urban areas, a lot of food is available for coyotes, so it seems they don't need to eat pets. But it's important to be aware of coyotes in your neighborhood, especially between December and March. If you have a yard and want to try and keep coyotes out, fix coyote rollers along the top of the fence so the animals can't get a purchase to jump in. Secure fencing (including a roof) is another option.

Of course, as these results show, cats are not just prey but also predators. Although cats mainly predate on mice, they will catch birds, reptiles, amphibians, insects such as crickets, and small mammals such as rats, voles, and rabbits, so some people prefer to keep their cat indoors to stop them from catching any prey. How much do pet cats actually threaten wildlife such as mice and birds? The answer is not straightforward. Many of the newspaper headlines we see about how many birds cats kill each year are based on studies of feral cats, and so do not apply to pet cats who are well fed. Some of these studies also have issues with methodology and extrapolate from very small samples. As well, people seem to feel more outrage about cats' predation on birds than other causes of declining bird populations, such as climate change, the collapse in the insect population, the use of insecticides and rodenticides, the reflective windows of skyscrapers, and so on. There's no evidence that cats are responsible for declines in bird populations.[10] They mostly kill sick birds, and they also kill rats and mice that might prey on bird eggs. When looking at studies of pet cats, there are some potential biases. Cats do not bring all of their catches home, and so surveys of cat owners may be an underestimate. On the other hand, people whose cats are good at catching things might be more likely to take part in such studies, making such surveys an overestimate.

To get around some of these methodological problems, one study near the Albany Pine Bush Preserve in New York used a combination of surveys, radio-collar tracking, behavioral observations, and scent stations set up within the preserve.[11] Based on owners' reports, pet cats who were allowed outdoor access caught an average of 1.67 prey a month, but based on observations, the researchers estimated the number was closer to 5.54 prey per month in the summer. The most common prey, by far, was mice (47% of the kills), with shrews (15%), cottontail rabbits (8%), and chipmunks (8%) caught in much smaller numbers. Birds made up 13.6 percent of the prey. As with the study of feral cats mentioned earlier, the pet cats in this study did not go very far into the forest. And despite the numbers of small mammals caught, the scientists did not think this affected their overall populations at all, especially given that they mostly caught young animals.

CONSIDERATIONS FOR SAFE OUTDOOR ACCESS

IF YOU WANT to stop your cat from hunting but still want to give them outdoor access, there are a few things you can do (on top of continuing to feed your cat so they don't have to hunt for food, of course). One is to attach a bell to your cat's collar. The little jangling noise means the cat cannot creep up on prey unannounced. Another option is the CatBib, a shield of flexible neoprene that dangles down from the cat's collar. Although the cat is still able to move around freely, including climb trees, the bib interferes with their ability to stalk precisely and pounce upon prey. When sixty-three cats in Australia wore either a CatBib or a CatBib plus bell for three weeks, they caught fewer birds and mammals compared with when they did not wear either.[12] An additional option is a circular collar cover called the Birdsbesafe, which is brightly colored and designed to be visible to prey that have good color vision, such as birds, reptiles, and amphibians, but not to prey that don't, namely mammals such as mice. In another Australian study, the rainbow-colored Birdsbesafe was more effective than the other colored versions and did reduce predation on birds but not mammals.[13] So this collar cover could be an option for people who want to protect birds but not mice. Most cats will tolerate any of these options.

Another option is outdoor access via a catio (enclosed patio) or otherwise safely enclosed area. A study in the UK found that cat guardians who installed such a system rated their pet's welfare as better, even for cats who had previously roamed freely. As well, the number of other cats coming by decreased significantly, which in itself may reduce or remove a potential source of feline stress. Cats with outdoor access often "timeshare" an area and avoid seeing each other. Dr. Luciana Santos de Assis, postdoctoral scientist at the University of Lincoln and first author of the study, says this result shows that "owners do not need to choose between strictly indoors or unsupervised access to outside to make their cats happy and safe. They can keep a high standard level of welfare with outdoor access and with a much reduced risk, as well as protecting the wildlife."

Many people are reluctant to restrict their cat's outdoor access because they believe going outside is good for their cat. One group of scientists wondered if instead they could satisfy a cat's motivation to hunt in other ways.[14] For the study, published in *Current Biology*, they recruited cats who regularly caught prey and divided them into groups to test the effectiveness of a Birdsbesafe collar, a bell, a daily five-to-ten-minute play session, a higher-quality food (that contained no grains, meat meal, or rendered meat), a food-puzzle toy, and a control group. The bell was not effective, as cats simply adapted their hunting technique. The Birdsbesafe collar led to a 42 percent reduction in bird catches but had no effect on cats catching small mammals, as expected. The daily play session, which was usually in the evening (when cats typically chase mice and other small mammals), resulted in 35 percent fewer catches of mammals but no significant effect on birds; it is not known if a morning playtime would affect bird catches. And the change in food led to 44 percent fewer bird catches and 33 percent fewer mammals. Surprisingly, the food-puzzle toy led to an increase in catching mice (but not birds). It is possible that the cats were frustrated or the toys were not introduced appropriately; more research is needed. This study suggests that if you want to stop your cat from hunting, feeding a higher-quality food and increasing playtime are likely to be beneficial.

If you decide to allow your cat access to your garden or a catio, think about how to make it a nice environment for your cat. "A garden where you just have grass, a lawn that is just mown, and you don't have lots of different soil or trees or bushes, it's not very interesting for the cat," says Dr. Kasbaoui. She told me about someone who turned a secondary one-room building on their property into a cathouse their cats could access via a cat flap; this could be done with a garden shed. But you can make a garden, catio, or balcony more interesting for your cat without spending a lot of money. Try to provide a range of trees, shrubs, and potted plants, as well as a shelter where the cat is safe and can take a nap. Different soil types and an outdoor litter box can also help.

Cats should only go outside during the day. If your cat is microchipped, try a microchip cat flap that will allow your cat(s) in but prevent other cats and raccoons from gaining access. Some models come with a timer so you can limit the hours your cat(s) can use the cat flap. It's a good idea for cats to wear a breakaway collar (one the cat can get out of if they get stuck on something) with a name or phone number printed on it. You also need to ensure your cat has permanent identification, which will most likely be a microchip, although it could be a tattoo. Whenever you move house or if you change a phone number, remember to update the microchip company (or your vet in the case of a tattoo). And if you find a cat who you think is lost, get them checked for a microchip at your local animal shelter or veterinarian before assuming the cat doesn't belong to anyone. Identification is still important for indoors-only cats in case they ever get out of the house.

If you are keeping your cat indoors only, you have to expect to work harder as a cat guardian to ensure that their needs are met. Options include making time for play, providing an environment with elevated perches, ensuring windows are safely ajar or screened so cats can smell the outdoor air, providing a cat running wheel (and training your cat to use it) for exercise, training your cat to go on leash walks, and providing a safe catio for your cat.

Veterinary behaviorist Dr. Karen van Haaften told me, "My big issue is indoor cats and the lack of enrichment and variety in the life of indoor cats. I think that's a major welfare problem that most cat owners don't pay enough attention to. And that stress of having a lack of enrichment in the home environment does manifest itself in physical illnesses too. So as a veterinarian, that's something that is very important to me. Like if somebody is bringing a cat into their life, it's not just an ornament that's going to sit in their house and look pretty. It's a living creature that has behavioral needs that have to be met. The bowl of food that's on the ground that's always full of the same boring kibble, that is a welfare issue for me." See chapter 10 for more on food-puzzle toys and how to feed your cat.

"One thing that would make the world better for cats would be for us to better understand what a 'cat' is. Cats are complex creatures which have co-existed with us for thousands of years. Yet society is still playing catch-up with regard to truly understanding their natural drives and needs. Compromised welfare is often the result of us misunderstanding their behavior and the ways in which cats communicate. Cats are instinctive hunters, territorial, self-reliant and yet highly adaptable creatures.

Often our modern ways of living mean compromising on a cat's natural behaviors. For example, keeping indoor cats without access to appropriate space or outlet for their instinctive need to hunt. One of the biggest causes of stress for cats results from the common misconception by humans that they need the 'company' of other cats. Cats are self-reliant and highly territorial. They have no biological requirement for companionship, especially from their own kind, as they are adapted to hunt and defend their territories alone. Consequently, the introduction of another cat into their territory can often be a highly stressful experience. Having a better understanding of our cats' real needs and what truly drives their behavior could help owners to provide cats with improved environmental and social conditions. In turn, this will help reduce problems and enhance the welfare and quality of life for pet cats."

—ELIZABETH WARING, cat behaviour course organiser, International Cat Care

TO SLEEP, PERCHANCE TO DREAM OF A MOUSE

AT NIGHT, HARLEY and Melina like to sleep on our bed. Harley has a specific routine, because he is a cat of routines. Both cats get a night-time treat: Melina on the antique nightstand that lives in the dining room, and Harley on a towel that is his and lives on the bedroom floor. He is always a bit slow to eat his, which is why Melina gets her treat in another room so she can't steal Harley's. Then we go to bed, and Harley jumps up on my side, tramples heavily over my pillow, treads on my hair, ignores me completely, and goes to see my husband. I used to be offended, but now I accept it. Harley goes to hang out on my husband's side of the bed where he gets some petting, purrs a lot, and sometimes tries to bite. Then after a while he goes and curls up next to or on my feet, where he will spend the night. If anything goes amiss with Harley's routine, he wanders around the bedroom and hallway howling as if something is wrong and he doesn't know how to put it right. At these times, sometimes if I pat the bed he will get the message. But other times he ignores me and then I end up getting up, picking him up, and putting him on the bed so that we can all sleep.

At some point after I've already fallen asleep, Melina comes to bed. Sometimes she wakes me by jumping up and landing with a thump (on a couple of occasions on my tummy). Other times she leaps over Harley. Then she settles down and becomes an immovable force. Whereas when she sits on my lap she is flighty and will leap off if I so much as move a muscle, on the bed at night she will not shift for anything, no matter how hard I try to pull the covers up or how much I fidget. But she does come and go, just for her own reasons, not because she's picked a spot where I'll get a cramp in my arm or where she gets all the covers.

Both cats also sleep during the day. Melina typically chooses what used to be our dog Bodger's bed, which, because she likes it, still lives under our bed. Harley snoozes for a while at the top of his cat tree in the dining room, except on a hot summer's day when he picks a sunny spot by the bedroom window and occasionally gets up to see what Melina is doing in the dog bed (he never tries to join her there, but sometimes he looks like he's contemplating it). Sometimes I wish I had the chance to sleep all day like that.

At night, the most common sleeping place for cats is their guardian's bed (34%), with 22 percent choosing furniture and 20 percent their own cat bed.[15] Your cat probably has favorite sleeping places around your home. It might be worth doing an audit to consider if there's anything you can do to make more spaces, or if your cat might like you to add a cozy fleece blanket here or a new cat bed there. Warming pads or heated beds can often be a favorite, especially through the colder months. Of course it's also possible that at some point you've bought what to you seems like a perfect cat bed only to have your cat turn their nose up and ignore it. In those instances you can try moving the bed to a better location, adding a towel or something else that already smells like your cat, or putting treats on the bed for your cat to find when foraging.

According to a 1984 paper that lists the overall sleep time and length of sleep cycles in many animals (including the three-toed sloth, golden hamster, and beluga whale), the domestic cat sleeps for 13.2 hours out of every twenty-four; that's more than half the day![16] As well, the length of time for which a cat sleeps at any one point is between fifty and 113 minutes.

Just like people, cats' sleep cycle has periods of REM (rapid eye movement) and non-REM sleep. REM sleep gets its name from the fact that the eyes can be seen moving rapidly, and this is the part of sleep in which dreams occur. When cats first fall asleep, they go into non-REM sleep, and can be woken up easily (for example, if they detect a sound).[17] After ten to thirty minutes, they enter REM sleep, which lasts for about ten minutes, and then they return to non-REM sleep. They will then move between periods of REM and non-REM sleep until they wake up. Up until about 6 weeks of age, kittens need more sleep than an adult cat. At 10 days old, all of a kitten's sleep is REM, but by 28 days this has dropped to about half of sleep time. As the kitten's activity levels increase, periods of non-REM sleep take up more of their sleeping time.

We know more than you might think about sleep in cats because a lot of early sleep research was actually conducted on cats. The late physiologist Prof. Michel Jouvet of the University of Lyon in France is responsible for discovering that during REM sleep (which he called paradoxical sleep), our muscles are paralyzed.[18] In one of his seminal studies, he took a cat, made a lesion to cause damage to a specific part of the brain called the pons, and then watched what happened when the cat was sleeping. During REM sleep, the cat moved around as if looking for and stalking prey and groomed, all while still asleep. This suggested that the cat was dreaming of those things, and that normally, muscle paralysis during sleep prevents the cat from acting this out. (Later, it was discovered that the same happens in humans.)

We used to think that dreams occur only during REM sleep, but we now know that they can happen during non-REM sleep as well. In a study that shed light on what happens in dreams, Prof. Matthew Wilson of MIT and a colleague ran rats through a maze and looked at the activity of neurons in the rat brains both while in the maze and later during REM sleep.[19] Amazingly, they found similar patterns of activity in the sleep phase compared with when rats were in the maze, suggesting the rats were running through what they had done in the maze as they slept. This reactivation of the neurons in sleep is thought to help consolidate memories. We know that sleep is important for the consolidation of memory in humans, and it is likely the same for cats too.

How do cats affect their humans' sleep? In a survey that asked women about their sleep partners and their sleep quality, dogs were said to be the best sleep partners. I mentioned that Melina sometimes disturbs my sleep when coming and going from the bed. This may not be unusual, according to this research from Dr. Christy Hoffman at Canisius College, New York.[20] Perhaps the good news for cats is that they were no more or less disruptive than a human sleep partner. Only 21 percent of cat owners said they slept better if their cat was touching them, whereas the remainder either did not mind or said they slept better if the cat was not touching them (38%). We know the main period when cats sleep is between noon and 6 p.m., and most owners said they thought their cat spent less than half the night on the bed (although 10% of people did not know). It's also worth knowing that cats can pretend to be asleep when they are stressed. Although this behavior might be more commonly seen in a shelter cat, it's something to keep in mind, especially during those first few days when you've just brought a cat home.

We do know that when cats live with people, they can adjust their schedules. In a study of ten cats who were given activity monitors to wear, cats who were routinely shut out of the house overnight were most active during the night-time.[21] In contrast, cats that slept indoors at night were most active during daytime, especially at the times when their owner was home and interacting with them. This ability to learn how to fit with our schedules is just one example of how cats can learn, which is the subject of the next chapter.

APPLY THE SCIENCE AT HOME

- Make a decision about outdoor access based on what is right for your particular cat and the environment where you live. Even if allowing outdoor access, it's best to keep cats in at night (which may mean calling them in before dusk).

- To give an indoor cat some access to the outdoors, consider ways to enclose all or part of your yard or balcony, have a catio, or train your cat to walk on leash.

- If there are times when you are home more often than usual, ensure your cat still has a quiet place to sleep during the daytime if they want to.

5

HOW TO TRAIN A CAT

· · · · · · · · · · · · · · · ·

S OME OF THE most fun I have had with Melina was teaching
her to "sit pretty"; in other words, to sit up on her haunches
with her front paws in the air. It felt slightly dangerous,
because I began by luring her with a treat and she was rather bitey
when trying to take it. She's not used to taking food from a person's hand in the same way that a dog would be. But these training
sessions were fun for Melina as well as for me. We did only a few
trials each day, but she would come running for them. I gave her
half a cat treat for each successful trial, which to be honest was
probably a bit much, but it wasn't easy to break the treats into
quarters and she really loved these treats.

I had already taught her to sit for a hand signal, and from sit I
used a treat to lure her backwards and up so that she was sitting
on her haunches. Once she had the hang of that, I began to make
the same movement but without a treat in my fingers. I wasn't

cheating her—as soon as she sat up, I gave her the treat. Although I'm generally a fan of feeding animals in the position you are training them to reach, I gave up on this strategy for the sake of not getting my fingers bitten, and instead put the treat on the floor next to her. Then it was time to try again. It actually didn't take as long as I thought it would for her to learn this position. Finally it was time to start saying "Sit pretty" before making the hand signal so that she would learn the verbal phrase. This part took longer, although I know she pays attention to the words I use. After a long period without practice, she will still sit pretty, but the hand signal is the bit she remembers best—which is perhaps not surprising as that got more practice than the verbal command.

What I also taught Harley and Melina in a far more systematic way than I had taught previous cats was to go in their cat carrier. The training had two aims: simply to teach them to love their carrier so that it would be a nice place to relax, and also so that they would go in it when I said "basket." It makes going to the vet so much easier, and I think all cats should be trained to do it. We'll get to what the science tells us about the benefits of training cats later in the chapter, but first it is important to understand how cats learn.

HOW CATS LEARN

AS A SOCIETY, we don't think too often about training cats, and indeed many people probably think of them as untrainable. But this is far from the truth. Cats learn all the time from their interactions with us whether we'd like them to or not. For example, they learn that when they come to sit on our lap we'll pet them, and depending on whether or not they like that, they will come to our lap more or less often. They learn that the shake of the treat packet means we'll give them a treat. And they learn very quickly that the cat carrier means an unpleasant trip to the vet.

Although teaching tricks can be fun and a nice bonding exercise for you and your cat, the most important thing is to train some key life skills, such as how to go in the carrier, how to be examined at the vet, how to be brushed and have their teeth cleaned, and to come when called. Dr. Sarah Ellis, coauthor of *The Trainable Cat* and head of cat advocacy at International Cat Care, told me that when training cats we are "teaching the cat the key skills that they need to live in society with us. And without those skills they often struggle. They're skills that are completely within the reach of a cat, you know. We're not asking for things that actually destroy the essence of what a cat is."

To train a cat, you need to have something your cat likes. Although it would be nice if a cat would do something just to be told "Good cat!" it doesn't work that way (it doesn't really work that way for dogs, either, who are much more used to being trained). Some cats who like being brushed will work for that—Harley is just such a cat who will come running as soon as I say "Brush!"—but for most cats, food makes the best reward. "There isn't that need to please," says Dr. Ellis, "so we have to think about what really is rewarding for a cat, because it's certainly not our social attention for most cats. And when we first start training a cat that's not been trained before, the most rewarding thing generally for cats is food."

Of course, being overweight is an issue for many cats (see chapter 10), so it's important to use only small rewards and to take account of the calories from training when feeding your cat. Types of food to use as rewards include little bits of tuna or prawn, pieces of cat treats, or little bits of wet cat treats (available in tubes). Regarding the size of treats, Dr. Ellis has this advice:

"Many people think about the size of the food reward that they give, and it is so often far too big. Because we sort of think in dog terms or even in human terms, and even the size that commercial cat treats come in are far too big to be a single training treat. So I very often recommend that if you are using commercial cat treats, use the freeze-dried ones or the semimoist ones, because you can pull them into much, much smaller parts. If we're thinking about a prawn, not a king prawn, just an average normal prawn, I would break that maybe into four or five parts at least."

Many cats are not used to taking treats from your hand and, like Melina, may try to bite or accidentally bite. To save your fingers, you may prefer to put the treat on a spoon or a little wooden stick (like a lollipop stick) or to offer wet treats from a dish or a tube. It's best to work in short stages, so the cat does not get bored or tired, and keep the level easy enough that they don't get frustrated. Especially in the beginning, this often means working in slower increments than you expect, and for short periods of time like five minutes. Your cat may also want a break between each trial; if they spend it purring and rubbing their head on you or the surroundings, that's a nice sign that they're happy. If they choose to walk off, that is of course their choice. Try again another time, and consider trying a better treat.

Cats learn in several ways, but the main ways that we use in training are types of associative learning: learning by consequence and learning by association with events.[1]

Associative learning

Operant conditioning means learning by consequence when the cat is either reinforced or punished for the behavior they just did. Reinforcement makes the behavior continue or increase in frequency, and punishment makes the behavior decrease in frequency. And there are two types of each, depending on whether the consequence was something added (positive) or removed (negative).

Positive reinforcement is the most well known and involves giving the cat a nice reward for a behavior so that they are more likely to do it again. Food makes a great reward for training cats. Negative reinforcement means that something unpleasant is removed when the cat does the behavior and as a consequence the behavior increases. This approach is not recommended in animal training because of risks to the animal's welfare. Fortunately it is rarely used with cats; however, you may have seen or heard of it in dog training. One example is when someone pushes the dog's bottom down and releases it when they are in a sit position in order to teach them to sit. The unpleasant sensation of the bottom being pushed ceases when the dog does the desired behavior, making the dog do the behavior more often. But, for dogs and cats, positive reinforcement works very well to achieve the desired behavior, so there is no need to use a negative reinforcement approach.

Negative punishment means that something the cat likes is withheld in order to decrease the frequency of a behavior. Maybe you are petting the cat and the cat bites you out of excitement, so you stop petting them until they stop biting to decrease the biting behavior. But notice I said out of excitement (i.e., they want more), because it's also possible (even likely) that the cat is biting because they don't want to be petted anymore. In this case they are applying positive punishment to you: something unpleasant has been added (a bite) to decrease a behavior the cat does not like (petting going on for too long). One example of using positive punishment with cats is when someone sprays the cat with a water bottle to stop them from going on the kitchen counter. This is not a good idea. A better approach might be to provide a nice high-up space, such as a cat tree near the counter, reinforce the cat with treats for using that space instead, and stop leaving food or toys on the counter that will tempt the cat to come and get them. For example, if you leave some fish on the counter and allow the cat to eat it, they've just been very handsomely reinforced to continue jumping on the counter.

If the consequences for the behavior stop happening, then the behavior will go away. This is called extinction. An example of when you might not want extinction to occur is if you have trained your cat to come when called by reinforcing the behavior with treats, but then you stop feeding the treats (maybe you ran out of treats and forgot to buy more, or maybe you thought the cat would work for nothing and come without the reward). Because the behavior is no longer being reinforced, the cat will stop coming when you call them. In the process, you might get something called an extinction burst, when the cat does the behavior again and again to try and get those consequences once more before giving up. Suppose the behavior is meowing at you for food, and you are trying to extinguish the meowing by ignoring it. You may find that an extinction burst can be very annoying!

Positive reinforcement and negative punishment are reward-based methods, whereas positive punishment and negative reinforcement are aversive methods. Some studies have shown that cats are more likely to have behavior problems when their owners use punishment. In one study, cats were twelve times more likely to eliminate outside the litter box in homes where their guardian used positive punishment.[2] In a study of cats who had been adopted from a shelter as kittens, aggression towards new people or objects, in new situations, and towards other animals in the home was much more likely when their guardian used positive punishment.[3] There has been a lot more research on dog training methods than on cat training methods, and the research with dogs shows that use of aversive methods has risks for animal welfare, including the risks of fear, stress, aggression, and a worse relationship with the owner. In addition there's some evidence that positive reinforcement is more effective as a training method than aversive methods.

These findings with dogs may apply to your cat too. For example, using a squirt bottle may startle your cat and cause them to feel fearful or stressed, and if they associate the squirt with you it may affect your relationship with them. If you punish your cat, it doesn't teach them what you would like them to do instead; and some behaviors (like scratching) are natural to cats, so it's up to you to provide appropriate ways for them to engage in that behavior (see chapter 1 on welfare needs).

Dr. Sarah Ellis told me, "I think to get people onside, they have to understand that punishment will—by the very nature it's called punishing a behavior—it will stop a behavior. But if that punishment is seen to be coming from you, you are also then perceived as punishing, therefore you are not perceived in a positive light. And therefore it can really damage the relationship that you have with that cat because for a punisher to really work it has to be really aversive. To stop that behavior it has to be stronger than the motivation to perform the behavior. Therefore if it's that aversive, and the cat associates you as being that aversive, you're going to really damage your relationship with that cat. Now, you're in your cat's life all the time: you feed the cat, you do other positive things with the cat, so you're suddenly giving this cat very, very ambiguous signals that 'sometimes I'm nice, and sometimes I'm not.' That can put the cat into a state of anxiety, and in extreme cases the cat could even begin to fear you. So you actually can then begin to create a situation where the cat hasn't just stopped performing the behavior you wanted it to stop performing, it's now actively avoiding you or actively fearful of you."

Of course, when teaching a behavior, you often need to break it down into small steps. There are a few ways to get the behavior you want. Capturing is when you wait for the behavior to happen naturally (like a sit), and then you say the cue ("Sit") and reinforce the behavior with a treat. This is a great way to pick up some behaviors, especially if your cat has a cute little behavior you want to see more of.

Another approach is luring, when you use a treat to lure the cat into position. For example, if you want to get your cat to sit, you can put a treat in front of their nose and lift the treat up and back. As the cat's head follows the treat, their butt will go down and they will sit. (You have to keep the treat close to their nose, because if it's too far away, they will probably stand on their hind legs instead to reach it.)

You can also use shaping, where you gradually shape a behavior through lots of little approximations. Typically, you would use a clicker or another marker (such as the click from a ballpoint pen or a word that you choose to use) to mark the specific point at which you see the behavior you want, and then reinforce that behavior with a treat.

Another type of associative learning is learning by association with events. For example, when your cat associates the cat carrier with an unpleasant car ride and an even more unpleasant vet visit and immediately runs to hide under the bed where they are out of reach, they are learning by association. This is known as classical or respondent conditioning, and the best-known example is Pavlov's dogs learning to salivate when hearing a bell. When physiologist Ivan Pavlov rang a bell and immediately gave the dogs food, the dogs would salivate in response to the food and soon began to salivate in response to the bell. Although Pavlov's experimental use of dogs to study digestion is distressing to read about, we will just focus on the mechanisms of classical conditioning here.

Some technical terminology goes along with this form of training, so let's switch to the example of using classical conditioning to teach that the cat carrier is not to be feared (counter-conditioning). We need something the cat really likes: let's say tuna. Tuna is known as the unconditional stimulus, and feeling happy about eating it is known as the unconditioned response—unconditioned because we don't have to do anything to make that happen; the cat likes tuna already. The cat carrier is the conditioned stimulus—the thing we want to train them to like—and liking it is known as the conditioned response. If we always bring the carrier out and then feed tuna, the cat will learn that the appearance of the carrier means she will be given something tasty to eat, and she will like to see the carrier because it predicts nice things.

If you're doing counter-conditioning with your cat, there are a couple of things to remember. One is that the scary thing (the cat carrier) must predict the treats, not the other way around. The other is that you need to have a one-to-one relationship in which the cat carrier always predicts the treats; if you get it out without offering the tuna, you will be undoing your training. And finally, you need to use really great treats.

If you want to try training your cat, you will find plans for teaching your cat to go in the cat carrier and to "sit pretty" at the end of the book. But there are some other ways in which cats learn too.

Other types of learning

You'll be familiar with single-event learning if you've ever become so drunk and hungover from one particular drink that you never want to have it again. As the name suggests, it means learning from something after it happens only once. In evolutionary terms, we can see this learning might be helpful to stop cats (or other animals) from continuing to eat poisonous substances. If that first trip to the vet in a cat carrier is awful for your cat, it's possible that will be a case of single-event learning.

Habituation is when your cat gets used to something that happens multiple times and doesn't really mean anything. For example, if your cat has been startling at the sound of the dishwasher but gets used to it and no longer flinches when the machine turns on, they are learning by habituation. In other words, they lose a behavioral response that was not learned (startling at the sound). This learning can happen with benign things like these noises made by the dishwasher or washing machine, which hopefully your cat habituated to as a kitten. The opposite of habituation is called sensitization, when an unlearned behavioral response (like startling at the sound of the dishwasher) gets worse and worse. This would be a sensible response if the dishwasher were dangerous, because the cat would learn to avoid it; however, since it's not dangerous, it just means the dishwasher becomes an unnecessary source of stress. Another example might be if you have a timid cat in a household with young children. The cat could habituate to the noises the kids make—and certainly many parents would hope this is the case—but it's also possible that instead the cat may sensitize and find those noises more and more frightening.

Cats continue to learn about other cats or people throughout their life, but this learning about the social world is especially important for kittens after the sensitive period for socialization. A range of positive experiences at this time will help them grow up to be confident and friendly. Other types of learning include paying attention to something because they see that you are paying attention to it (social facilitation), or paying attention to something like a toy because you or another cat are manipulating it (stimulus enhancement). Scientists found that kittens learn a task more quickly when they have seen an adult cat do it first, especially if the adult cat was their mother.[4] Cats, like other animals, also have some behaviors that all cats do and that they do not need to learn, which are known as modal action patterns. An example is the crouch-stalk-pounce used in hunting. But these behaviors can change through learning. Mother cats teach their kittens to hunt by bringing them prey they have already killed, and then, later, prey that is still alive for them to catch. At the time of writing, there is one example in the literature of a cat learning to copy what a person does in response to the command "Do as I do," although more research is needed to test this approach more thoroughly.[5]

"I think the one thing that would make the world better for cats at an individual level would be to improve opportunities for cat owners/guardians to train them. Give them more information about how unique each individual cat truly is but that they all can be trained, which could give them skills that would make coping with our anthropocentric environments much easier for them. What enrichment do they like? Do they prefer treats or toys for training?

At a societal level I think there are a few things that could change:

1. We need to rethink how we manage cats at a community level and in animal shelters. As long as those responsible for caring for cats continue to send the message that cats are second to dogs and we can put them in cages much too small for an acceptable life, the public won't see them differently.

2. We also need to reconsider how we deal with nonhuman animals, particularly cats, in our legal system. How we label them (are they companion animals, wildlife, feral...) matters in how we CAN treat them."

—MIRANDA K. WORKMAN, PhD, director of behavior and research, SPCA Serving Erie County

THE BENEFITS OF TRAINING CATS

AS WE'VE JUST seen, cats can learn, and training them can be very beneficial. "It's so important—training them, doing clicker training," says veterinary behaviorist Dr. Wailani Sung. "Teaching them basic things like a simple touch and go to your mat is so important. If [cat guardians] can start doing that, they will really develop a stronger bond with their cat."

One of the times when training can make the biggest difference is when you need to take your cat to the vet. Many cats have learned that the cat carrier predicts a trip to the vet and will resist being put in the carrier with tooth and claw. But scientists have shown that cats can be trained to like going into their carrier, and that it makes life much easier at the vet.

For a study published in *Applied Animal Behaviour Science*, scientists at the University of Veterinary Medicine in Vienna, Austria, tested twenty-two cats who lived at their laboratory.[6] They randomly allocated the cats either to a group that was trained to use the cat carrier or to a control group that wasn't. The training plan began with teaching the cat to go in the bottom of the carrier and built up to going in the carrier for a very short ride in a car (50–90 seconds). All of the cats had twenty-eight training sessions of about eight minutes each. The cats worked for tuna, meat sticks, or assorted cat biscuits, depending on their preference, and they earned about four treats a minute.

To move on to the next stage in the training, the cats had to either achieve the goal of that session or have had six sessions at that stage. Of the eleven cats in the training group, only three fully completed the training. A further six cats reached the final stage but did not finish it, and two cats reached the penultimate stage. Before and after the training, cats from this group and the control group were taken for a mock vet exam in which one of the scientists acted as the cat's owner while another played the role of the vet. The scientists looked at the cat's behavior while being put in the carrier, in the car, and during the vet exam, as well as their scores on a standardized stress scale called the Cat Stress Score.

Overall, the cats who received training to use the carrier were less stressed according to their Cat Stress Scores and behavior. The trained cats were less likely to pant or hide during the car ride, and some of them even ate treats during the journey. The vet exams for the trained cats were completed significantly more quickly than those for the untrained cats. The cats made fewer attempts to escape and spent less time hiding, and in most cases the vet exam could be completed in full (sometimes it ended early because the cat would not tolerate having rectal temperature taken). As well, the scientists used the kind of carrier with a top that comes off and found that cats liked to stay in the bottom of the carrier during the vet exam, showing that it was a safe space for them.

If these results inspire you to train your cat to use their carrier, you will find a training plan in the appendix (see page 244). Once your cat completes the training, it would be a good idea to do "reminder" sessions from time to time so that they continue to associate the carrier with good things. It's also a good idea to keep the carrier out, such as in the living room, where it can become a normal piece of cat furniture (rather than something that signals a trip to the vet) and where your cat can choose to relax (to continue to build positive associations). It has even been shown that cats can be trained to accept a blood draw at the vet.[7]

Other research on the benefits of training cats has looked at the effects of training on shelter cats. One study, published in *Animals*, looked at whether shelter cats could be taught tricks (to sit, spin, or high-five, and to nose-touch either the trainer's finger or a chopstick, depending on how fearful the cat was).[8] Over two weeks, each cat took part in fifteen five-minute clicker training sessions. By the end, 79 percent had learned to nose-touch the target, 60 percent to spin, 31 percent to high-five, and 27 percent to sit. Even very shy cats took part in the training (rather than choosing to hide at the back of their cage) and learned some tricks. This result shows that tricks training is suitable for any cat. As well, it seems likely that the training sessions helped shy cats learn positive associations with people.

Another study, published in *Preventive Veterinary Medicine*, looked at the effects of training shelter cats who are frustrated in that environment.[9] Some cats brought to the shelter will pace, chew the bars, put their paw through the bars, tip their food and/ or water out of the bowl, and so on. We might think they've had a wild party and trashed the place, but these are behaviors that show they are frustrated at being stuck in the cage or room they are in. Frustration is obviously a welfare issue for those cats, and so researchers at the BC SPCA tested whether or not a training program could help. Fortunately, frustration is not that common: out of 250 cats assessed for this research, only fifteen were found to be frustrated. The frustrated cats were randomly allocated either to a control condition (eight cats) or a training condition (the remaining seven cats) in which each cat was taken out of their cage four times a day to have a ten-minute training session in another room. Using food rewards and a clicker, they were taught to do a high-five in response to the cue "Give me five." Samples of poop were taken every day and analyzed for cortisol, a hormone that is a measure of arousal. The cats' behavior was monitored via video cameras in their cage, and the researcher also assessed the friendliness of the cats.

The results showed that training was good for the frustrated cats. They showed more signs of being content, such as normal grooming, lying on their side in a relaxed posture, rubbing their head or body on things in the cage, and spending more time sitting at the front of the cage. In contrast, over time the control group of cats became apathetic, typically after six days of trying to escape. They did not eat as much or groom themselves properly, and they spent a lot of time sleeping. Stool samples showed that cats in the training group had higher levels of immunoglobulin A, which can protect against upper respiratory infections. In line with this finding, the control group was significantly more likely to get an upper respiratory infection during the time of the study. The training activity involved time out from the cage and time spent with a human, as well as the training itself. Any of these, or the combination, might have caused these beneficial results, so more research about the role of training is needed.

If you want to give training a try, focus on learning that will make a difference to your cat's welfare (such as carrier training, toothbrushing, taking medication, and nail clipping) or provide cognitive enrichment for them. If you begin this training when you have a kitten, it will prevent them from developing the negative associations that many adult cats have. Keep sessions short and make them fun. Give your cat a choice of whether or not to participate (it's up to them if they choose to walk away!). And use something your cat loves as positive reinforcement to keep them motivated. Ideally, training sessions (even for things the cat needs to learn) will be fun and pleasant activities for both of you. And they can make a big difference to trips to the vet and for grooming, the subject of the next chapter.

APPLY THE SCIENCE AT HOME

- Think about the behaviors that it would be useful to teach your cat, such as coming when called, going in the cat basket, being groomed, having their teeth cleaned, and so on. Think life skills rather than tricks. Then follow a gradual plan to teach them.

- Identify the type(s) of food that will work best as positive reinforcement for your cat, and use it. Don't expect your cat to work for free.

- Tricks can be a fun activity for you and your cat and can provide both cognitive enrichment (the learning part) and food enrichment (the rewards part).

- Don't use punishment to train your cat. It could cause your cat to be stressed and to associate the punishment with you, affecting your relationship. As well, it does not teach your cat what you would instead like them to do.

- As well as training your cat about how you would like them to behave, make sure you are also meeting their needs. For example, you cannot expect to train a cat not to scratch, because scratching is a normal feline behavior, and opportunities to engage in normal behaviors are an important part of good feline welfare. But you can provide good scratching posts in locations your cat is likely to use them, and give positive reinforcement for scratching these posts.

6

THE VET AND GROOMING

. . . .

HARLEY AND MELINA are both short-haired cats and don't need much grooming, but because Harley sheds a lot, I decided to start brushing them occasionally. It turns out Harley absolutely loves to be brushed, so much so that clearly I should have started much sooner. Brushing him is now a daily ritual. He has a good sense of time, and he expects to be brushed at exactly the same time every single day. If I'm even a few minutes late to go and get the brush, he will come and find me and howl at me, and then run along the hallway with me, almost tripping me up. It takes only a few minutes to brush him, but it's clearly a few minutes of bliss from Harley's perspective.

While I'm brushing Harley, he rubs his head on the bedside table or on my hand. He likes being brushed along the stripe of fur on his back the best. He tolerates me brushing his trousers, but will sit down if it goes on too long. And while he's standing up I reach round to brush his belly. This would not be safe to do with him lying down, but he seems to enjoy being brushed while he stands up. I move the brush in short strokes, kind of similar to the short licks one cat would give another cat. And he purrs loudly all the while. If I stop too soon he comes and rubs on my hand or else howls at me for more.

Melina is less keen on being brushed. She tolerates it for a short time, but of course she's free to run away when she's had enough, and she will if it goes on for too long. I brush her only once or twice a week and try to stop while she's still happy, even if it's only a few strokes of the brush that day. All cats need some husbandry or vet trips in their life, but unfortunately it's something cats typically don't like.

WHY YOU SHOULD TAKE YOUR CAT TO THE VET (AND HOW TO MAKE THOSE VISITS EASIER)

TRIPS TO THE vet are necessary for all cats, to prevent illness (including vaccinations and parasite treatments) and to treat them if they get sick. Yet only 60 percent of cats (compared with 85% of dogs) had been to the vet in the previous twelve months, according to a survey published in *Journal of the American Veterinary Medical Association*.[1] Twenty-one percent had been to the vet in the last one to two years; 6 percent, three to four years ago; 8 percent, five or more years ago; and 7 percent of cats had never been or the owner did not know. Only 83 percent of cat owners had a veterinarian.

The three main reasons people did not take their cat to the vet included the cat struggling not to be put in the carrier, concerns about the costs of veterinary care, and not understanding the reasons for preventive veterinary visits. Amongst both dog and cat owners, 42 percent said they thought indoor pets did not need checkups, and 32 percent said routine checkups are unnecessary. More than half (58%) of cat owners said their pet hates going to the vet, and 38 percent said just thinking about going to the vet was stressful. These results show the need for education to explain why veterinary visits are so important, including routine vet care that may help to prevent medical issues from developing or worsening, and may prolong the life of your pet.

In fact, 56 percent of cat owners said they would go to the vet more often if they knew it would prevent problems and expensive treatment later, although this begs the question of why so many would still avoid it even with that knowledge. As well, 43 percent of cat owners said they would go to the vet more often if they were convinced it would help their cat live longer. Almost all (97%) vets think adult pets who aren't yet seniors should have an annual wellness exam, according to a follow-up survey reported in *Journal of the American Veterinary Medical Association*.[2] For senior pets, more than half of vets recommend a wellness exam twice a year, with the rest recommending once a year.

As mentioned in the previous chapter, training your cat to go in their carrier can make a big difference to your cat's experience at the vet. Having a carrier where the top can be removed, so that the vet can examine your cat in the bottom of the carrier, can also help reduce your cat's stress. And of course it is best if kittens have happy, positive visits to the veterinarian, as these will help to set them up for future happy visits as adults.

There is often a mismatch between what pet guardians expect at the annual preventive care consult and what veterinarians think this consult is for, according to research published in *Veterinary Sciences*.[3] The preventive care consult is not just for vaccinations; it is also for an overall health check and for parasite control. Pet owners expected that their pet would have a health check, as well as any vaccinations that were needed, and that this visit would be reassuring for them. People with more experience of owning a pet did not expect their vet to find anything they did not already know about. Most of the pet guardians who took part in this study were dog owners, and so more research is needed to find out any specific expectations or concerns of cat guardians. For older pets, vets expected to find medical issues. The most common issues amongst cats were weight loss, dental disease, and excessive thirst (which is often associated with diabetes). The authors of the study say that the preventive consult should be not just about vaccinations and parasite control, but also about preventing obesity, dental problems, and behavior issues.

Most cats find vet visits stressful. Amongst the main reasons are that cats are not used to going places, the vet's office smells of disinfectants, they are handled in ways that they don't enjoy, and there is nowhere to hide. Traditionally, full restraint for cats at the vet has involved them being held on their side with their neck and legs immobilized. When scientists tested shelter cats with either this full-body restraint or a passive restraint that was gentler and allowed for some wiggle room, they saw many more signs of stress in the cats subjected to the full-body restraint.[4] The signs were more lip licks, faster breathing, dilated pupils, and the ears being back or to the side. Cats in full-body restraint were 8.2 times more likely to struggle, and then 6.2 times faster to leave the exam table when released compared with those given the passive restraint. Signs of stress are also seen when cats are scruffed (held by the scruff of the neck), so scientists recommend avoiding scruffing cats.[5] Full-body restraint is still common, but less likely in Canadian than in US vet practices, and less likely in practices certified by the American Association of Feline Practitioners.[6]

A survey in Italy reported in *Journal of Applied Animal Welfare Science* found that 24 percent of cats had bitten or scratched their guardian at the vet and most had been aggressive to the vet at some point in their life.[7] Only a third of cats would let the vet touch them anywhere, and for many cats the tummy, tail, and genital area were reported as off-limits. A third of cats objected to having their temperature taken, and a third to having vaccinations, while around a quarter had refused a blood sample. Many vets were using food to try to make the experience better, and 78 percent of the cats had been offered food at the vet. Half of cats (49%) refused the food that was offered, and 29 percent were said to be suspicious of it. These results show that using food helps cats be less stressed at the vet, but it's essential to help cats feel relaxed enough to be able to eat it.

Dr. Chiara Mariti, veterinary behaviorist and coauthor of the study, says, "I would suggest familiarizing kittens with manipulations in a gentle, gradual, and progressive way, associating any handling with positive emotions and stimuli. Also positive associations with anything related with the travel, especially the carrier, can help; the appropriate use of pheromones may be beneficial. I would stress the importance of avoiding the association of the carrier with the visit to a vet clinic. Some vets suggest the use of towels to gently 'wrap' the cat in; it seems to calm the cat during the visit and to reduce the need for physical restraint.

"Owners should try going to the clinic with an appointment, in order to avoid long stays in the waiting room (usually the car is better). When getting to the clinic, they should avoid contact with other animals, and if this is unavoidable, they should put the carrier as high as possible (shelves, chairs) to give the cat the opportunity to feel a bit safer. Visits to the clinic as a kitten, just to familiarize them with the place and the vet, should be encouraged."

If your cat did not get positive vet experiences as a kitten, or even if they did but have subsequently learned to dislike their carrier and vet visits, you can work on changing that. Always take this kind of training very slowly because you can only go at the cat's pace, which is probably much slower than you would like. Trying to go faster can often make things worse. But while you are working on teaching your cat to like their carrier and being handled, they may still need to visit the vet. If the experience is likely to be highly stressful for the cat, a vet will sometimes recommend the use of psychoactive medication before the visit.

Veterinary behaviorist Dr. Karen van Haaften told me that sometimes the cat is so terrified they freeze, and that's a welfare concern. "There are a lot of cats that don't have dangerous behavior at the vet clinic," she said, "but they go into that freeze mode, so it's actually very easy for a vet to do an exam and get what they need to do done. It's not even a behavior problem from anybody's point of view. But from the cat's point of view, that's an extraordinarily stressful event that is causing that cat to have such a severe lack of behavior and not be themselves at all. So I would recommend giving that animal a kind of anxiolytic before the vet appointment to help them cope with that stress, even if it's not a behavior concern necessarily."

"People often think of cats as 'low maintenance' pets compared with dogs, but that's a misconception, and unfortunately can lead to a more laissez-faire approach to their care. I would love to see cats exercised more and kept lean. Fat cats are much more prone to diabetes, hepatic lipidosis, and arthritis.

Many cats, especially indoor-only cats, eat out of boredom—so ditch the bowl and use food toys! Urinary problems are very common, so many cats benefit from increasing water intake by offering wet food, water fountains, and extra water bowls. Dental problems are very common and often painful, so I recommend toothbrushing and being prepared to start preventive dental care early on in life. Cleanings and treatments under anesthesia are often necessary to keep the mouth healthy and pain-free.

Cats are great at hiding symptoms of medical problems and it's easy to overlook many of the signs they do exhibit. Weight loss, decreased interaction and engagement with their usual activities (play, jumping, grooming...), changes in appetite or water intake (up or down!), and changes in potty habits (frequency, location, size/consistency) are all signs that should be investigated by your vet sooner rather than later. Subtle changes can be very significant, and regular wellness exams and blood screening can help catch changes that the cat has successfully hidden from us."

—**DR. RACHEL SZUMEL,** small-animal veterinarian in South Lake Tahoe, California

CHOOSING A VET

MAKING CHANGES TO the veterinary waiting and exam rooms, using considerate handling techniques, and tailoring the exam to the individual cat can all help to reduce stress.[8] Several programs can help you find a vet who is trained to keep your cat's stress levels as low as possible. The Cat Friendly Practice program was developed by the American Association of Feline Practitioners and the International Society of Feline Medicine[9] to help cat guardians choose feline-friendly veterinary practices and clinics anywhere in the world. Vets that follow this program can suggest ways to reduce stress before the visit, will ensure you have a calming waiting room and a cat-only exam room at the clinic, and are well versed in recognizing signs of pain and illness in cats.

Another program that aims to reduce stress for animals at the vet is Fear Free, which was started by Dr. Marty Becker and has grown to include not just individually certified veterinarians and vet techs but also veterinary practices. Dr. Becker told me, "I graduated from veterinary school being taught that animals didn't feel pain. So I graduated 1980, forty years ago, literally told by professors in neurology and in clinical medicine that pets didn't feel pain, and if they did it was good, because they would be inactive and not tear the stitches out or walk on the leg we just fixed. And I'm thinking, how in the hell could I have thought that, when we dehorn and brand a cow and they literally bellow, just scream. That's not pain? Or you step on your dog's foot and they cry? I mean how dumb was that? I think it's just the fact that pets have a broad range of emotions that we need to recognize." Fear Free was developed in response to the realization that stress and anxiety at the vet is bad for pets. Fear Free–certified veterinarians and vet techs have training in how to deal with fearful and anxious pets and how to protect pets' emotional well-being, and are required to complete continuing education to maintain that certification.

Another certification that some vets and other professionals might have is Low Stress Handling. This practice is based on the late Dr. Sophia Yin's methods for reducing stress when handling dogs and cats for veterinary procedures. For example, Dr. Yin developed several techniques for wrapping a cat in a towel to use minimal restraint, and reduce harm to themselves and the vet if they become stressed.

Many pet guardians are concerned about the cost of veterinary care. Dr. Adrian Walton of Dewdney Animal Hospital in Maple Ridge, BC, told me that improvements in technology and techniques have dramatically changed the vet experience for the better. He explains the process if your cat has surgery: "Your cat is given a premed anesthetic, heart protectant, and pain control before we do anything; they're sedated, intubated, they're monitored by a trained veterinary technician, they're done in a sterile surgical suite with sterilized equipment, fully gowned veterinarian, IV catheters, heart monitoring, worming, basically the same quality of medicine as you would get if you went into surgery." Yet, he says, some vets still use practices that were common in 1988 when he started working in a vet clinic. Cats got much less individual care and monitoring, he says. "If you're wondering why there is a different price between different vets, ask some questions, find out what the difference is. The vet should be able to explain." In other words, price may reflect different levels of service.

Other factors to take into account when choosing a vet are the opening hours (including whether they are open in the evening or at weekends), whether they take emergencies, and whether you will see the same vet or a different vet at each visit. If you're not happy with your vet, change to another one. As well, note where the emergency vet clinics are in your area and keep their phone numbers at hand just in case.

VACCINATIONS, PREVENTIVE MEDICINE, AND GROOMING

YOUR KITTEN'S FIRST experience at the vet will likely be for vaccinations and deworming. Vaccination is an important way to protect pet cats from disease. Kittens get some natural protection from disease due to antibodies from their mother in the colostrum, the milky fluid released by the mother's mammary glands just after birth. These antibodies protect the kittens in the first few weeks of life and then that immunity drops off. Since these antibodies can prevent vaccines from taking properly in young kittens, it's a case of finding the right timing to receive early vaccinations. Like other young animals, kittens are very susceptible to disease until they receive their shots. Guidelines from the American Animal Hospital Association (AAHA) and the American Association of Feline Practitioners (AAFP) suggest vaccinating kittens from 4 weeks of age with boosters at gaps of 2-4 weeks up until 16-18 weeks to ensure protection.[10] For the best protection, they recommend a final vaccine for FHV-1 (feline herpesvirus type 1), FCV (feline calicivirus), and FPV (feline panleukopenia virus) at 6 months of age, even though this means an extra visit for the final course of the rabies and FeLV (feline leukemia virus) vaccines at 1 year. (It was previously recommended that all the final vaccines occur at 1 year of age.)

The AAHA and AAFP guidelines about vaccinating cats divide vaccines into core (recommended for all cats), noncore (recommended for some cats), and one that is not recommended because the benefits are uncertain (for feline infectious peritonitis). The core vaccines are for feline herpesvirus type 1, feline calicivirus (cat flu), feline panleukopenia, rabies, and feline leukemia (for cats under 1 year). The noncore vaccines that your vet might recommend, depending on your cat's lifestyle and your location, are feline leukemia for cats over 1 year old, *Chlamydia felis*, and *Bordetella bronchiseptica*. "From the veterinarian point of view, I wouldn't treat a cat that is indoors and a cat that is outdoors the same," says Dr. Naïma Kasbaoui. She suggests more preventive treatment for cats with outdoor access. "For example, I would be much more strict on deworming, anti-flea treatment and also maybe, depending on the environment, also on vaccination. For me, we were educated that when you have an indoor cat you do panleukopenia and cat flu because those are the things that you can bring on your shoes. But if your cat is going out, maybe feline leukemia virus too." Your vet will help you decide if these vaccinations are a good idea for your cat or not.

Vaccinations are generally recommended for all cats, even indoor cats. They still have the risk of getting sick, by going outside (on catios, on patios, for on-leash walks, or by escaping), from other animals in the household, or from bacteria, viruses, or fungi coming in on the clothing or belongings of household members. Rabies vaccination is required by law in some areas.

The risk of side effects from vaccination is low. Tumors arising at the injection site—called injection-site sarcomas—are rare but can occur, and this is why your vet will let you know where the injections were done if you are not present for them.

Parasite control (deworming and anti-flea treatment) "depends on the cat," says Dr. Walton. "If your cat is actively hunting and moving about outdoors, a monthly dewormer is recommended. If they sedately sit in your house and catch the occasional fly, probably not as frequently. It also depends on the owner's lifestyle. Immunosuppressed? Cancer, AIDS patient? Young children? More commonly."

Grooming is another important part of caring for your cat. Many cats with long fur, such as Persians, need to be brushed every day as they are not able to properly groom themselves. Don't get a long-haired cat if you're not going to keep up with daily brushing, or they will end up with matted fur that gives them pain. Short-haired cats can often take care of their own grooming needs, but you may still wish to groom them sometimes. When cats groom themselves, they swallow old and dead hairs, most of which pass through their digestive tract with no problems. But sometimes those hairs form a hair ball that cats will then throw up. This is normal from time to time, but if it happens often you should speak to your vet in case there is a medical issue or stress. Similarly if you notice bald spots in your cat's fur, speak to your vet.

All cats, even short-haired ones, may need help with grooming as they get older. The trousers and tummy are particularly in need of extra help from you, but may not be areas that your cat likes to have brushed. It is best to get them used to being brushed as kittens, when they are much more likely to tolerate the process, but adult cats can be taught to like it so long as you work slowly.

Try to make grooming a positive experience, and use treats and head-rubs to help your cat enjoy it. Move the brush in short strokes, and start somewhere your cat is likely to find pleasant, such as on the sides of the chest and shoulders. Always stop while your cat is still enjoying the process. Don't think you'll just carry on and finish a certain bit, because if the cat gets fed up, they may not be so keen to let you try brushing them again. If mats appear, you can often gently pull the bits of fur apart with your hands. This is preferable to cutting them off, which carries the risk of accidentally cutting your cat. For bigger mats, or for cats who won't let you groom them, your veterinarian or a local groomer will be able to help. In some cases the cat may need to be sedated to allow it to happen; your veterinarian will talk with you about the risks and benefits.

Some types of cat (especially those with flat faces such as Persians and Exotic Shorthairs) will need to have their eyes and face washed regularly. Do not bathe your cat unless it is essential (such as after diarrhea), because it will strip the oils from their coat. And never trim your cat's whiskers, because they are full of nerve endings. Your cat relies on their whiskers to judge space, to show how they are feeling, and (if they catch prey) to help with the catch.

Cats that spend time outdoors may be able to keep their claws in good condition, but indoors-only cats and older cats may need their nails clipped from time to time. Again, ideally you will get them used to nail clipping as kittens, but it is quite possible to teach older cats to allow it by working through a gradual training plan. If you don't like to use the curved and blunted nail clippers designed for cats, try a nail file instead. It is important to avoid cutting the quick, the part at the top of the nail that contains blood vessels, nerves, and other tissues, as this will be painful and cause bleeding.

Most cats are good at grooming themselves, but some will need a little help, especially as they age. *JEAN BALLARD*

Your vet or groomer will be able to trim your cat's nails for you if you don't fancy trying it yourself. And remember to always have good scratching posts available (see chapter 11) to help cats keep their claws in tip-top condition; after all, scratching is a normal behavior for cats.

Caring for your cat's teeth is also important to prevent pain and inflammation, and good dentition is good for your pet's overall health as well.[11] In people, dental issues are related to general health, particularly vascular disease, as inflammation and bacteria can spread through the body. Although it is not yet known if the same is true of cats, pain and inflammation can definitely be associated with issues with cats' teeth. Signs of dental problems include bad breath, sores in the cat's mouth, red or bleeding gums, drooling, and a poor appetite, as the cat is reluctant to eat.

When cats were given a free vet check at a vet clinic in Belgium, dental disease was the most common condition after overweight and obesity, affecting 21 percent of cats.[12] Meanwhile, amongst a random sample of cats attending the vet in the UK, as published in *The Veterinary Journal*, 13.9 percent had periodontal (gum) disease.[13] A study published in *Journal of Veterinary Internal Medicine* found that both moderate and severe dental disease are associated with later development of chronic kidney disease (CKD), a common health issue in older cats.[14] However, despite the time difference between the records of dental disease and the later CKD, it cannot be confirmed that dental disease is the cause, as other risk factors may also have played a role. Research on dogs published in *Journal of the American Animal Hospital Association* has shown that dental cleanings are associated with a 20 percent reduced risk of death.[15] This is another reason why vet visits are so important, even if your cat is not due for vaccinations.

Gradually teaching your cat to let you brush their teeth will help with their dental health. Dr. Rachel Szumel is a small-animal veterinarian in South Lake Tahoe, California, who has written a course on toothbrushing for dogs and cats. I asked her why it is important to brush your cat's teeth. "The quick answer is so you don't get stink-breath when you're trying to snuggle with your cat," she said. "The longer answer is for their overall health and well-being. Gingivitis and periodontal disease cause chronic inflammation and infection and that affects everything. We think it probably has a lot to do with kidney disease in older cats, which anybody who's had an older cat is probably familiar with. Keeping those tissues in the mouth healthy helps keep their whole body healthy. And they will feel a lot better. Also it keeps them a lot more comfortable."

Dr. Szumel says you should clean your cat's teeth every day. "Go at cat pace, which is slow," she says, "and find the right motivator. Don't expect your cat to do it for the love of toothbrushing. Expect them to do it for the love of sardine paste or baby food or whatever other stinky thing you can find that really floats your cat's boat. And then really gradually plan to introduce it, maybe brushing one tooth at a time at first or even just touching their face, lifting their lip, and then introducing the toothbrush. So that it never is overwhelming to them." Use a suitably sized toothbrush (cat toothbrushes are available) and toothpaste designed for cats. If you have multiple cats, use a different toothbrush for each cat. Although finger brushes are available, be careful not to get bitten.

Your vet will check your cat's teeth as part of the preventive medical visit. If dental cleaning is recommended, this can only be done by a licensed veterinarian with the pet under anesthetic; it cannot be done by dog groomers, as nonanesthetic cleaning does not include scaling; does not improve the pet's dental health; and can cause bleeding, pain, and infection, according to the *American Animal Hospital Association Dental Care Guidelines*.

If you get it right, training your cat to have their teeth brushed could be an enriching activity, because training should be a fun experience for the cat. In the next chapter, we'll look at different ways of providing enrichment.

APPLY THE SCIENCE AT HOME

- Train your cat to like their cat carrier, as it makes taking them to the vet easier and also helps make the veterinary exam easier.

- Choose a vet that will keep the cat's stress levels low by looking for a Cat Friendly Practice, a Fear Free–certified vet or practice, and/or a vet with the Low Stress Handling certification. Try to see the same vet each time so they get to know your cat.

- Vaccinate your cat as per your vet's recommendation, and take your adult cat to the vet at least annually (more often for kittens and senior cats).

- Get your cat used to being brushed, being groomed, and having their teeth brushed when they are a kitten if possible. Otherwise use a gradual plan to get them to like these activities. Work slowly, using good rewards like canned tuna or liquid cat treats, and always be careful to stay within the cat's comfort zone.

7

ENRICHMENT FOR CATS

. . . .

KEEP THE CATS' toys in a box under the coffee table. Every now and then I will pick up the toys from the floor, put them in the box, and pick out a few different ones to place around the room. The idea is that the cats won't get bored because the toys are always changing. But the box itself has become a source of enrichment for Melina. She knows that's where the toys are kept and will work hard to get the lid off. It takes her a few moments of effort to begin to lift the lid, and then she nudges it aside and peeks in. Very often she will nose around in the box, pick up a toy in her mouth, and go running across the room with it. She likes to choose her own toys.

We are used to hearing that you have to give dogs a job or they will find one of their own, but to some extent it's true for cats too. Sometimes I think Melina thinks her job is opening doors. She will go all along the hallway that runs down the middle of our house, starting with the door to the laundry chute. She puts her paw under the door and pulls it open. Then she moves to the linen cupboard and does it again. Then the coat cupboard. Harley, meanwhile, will only open doors to get somewhere or to get something. He is especially interested in the door to the cupboard where we used to keep Bodger's food, but he checks the other kitchen cupboards too. One time, I saw a cupboard door ajar and peeked in just in case there was a cat in there. Harley was curled up fast asleep inside the frying pan, which seemed to be the perfect size for him. I wish I had been able to take a photo before he woke up and moved on, miffed at being disturbed!

WHAT IS ENRICHMENT?

FOR ALL CATS, but especially indoor cats who are unable to find their own opportunities for learning and play outdoors, enrichment can be a good way to improve their welfare. According to a paper by Dr. Sarah Ellis in *Journal of Feline Medicine and Surgery*, "environmental enrichment is the process of adding one or more factors to an animal's environment in order to improve the physical/psychological welfare of the animal."[1]

Dr. Wailani Sung says people "need to give cats things to do. Their environmental enrichment is pretty lacking. And some people know, 'Okay I'm going to get them a catnip toy,' but that's about it. I always recommend having multiple feedings throughout the day, small feedings, because typically on average a cat spends maybe 45 to 60 percent of their day hunting for food, and there's a pattern of hunting, resting, grooming, and things like that, walking the territory, and spraying. Cats have the life of Riley, 'Here's our feeding trough, we stuff our face' and then what else is there for the cat to do? Not much, right?! 'This toy that

I batted around for the past five days, it's no longer very interesting.' So people need to rotate their toys. I have feeding strategies where we have puzzle toys, and sometimes I have people that feed their cats with canned food. I make them hunt for their food. We get three or four little ramekin containers and we divide the food up and put them at different locations throughout the house so that the cats learn to start hunting for their food. 'Okay it was here yesterday, where is it now? Oh 2 feet away or 4 feet away from the last location.'"

The two main types of enrichment are animate and inanimate. Animate enrichment involves social relationships with people or other animals the cat likes to spend time with. Relationships between cats and their people are covered in chapter 8, and between cats and other animals in chapter 9, so this chapter will focus on inanimate enrichment—enrichment involving objects, as well as general principles for using enrichment and assessing whether or not it is working. After all, there's no point in providing particular types of enrichment if your cat isn't interested—or worse, is scared of them.

One reason to use enrichment is to give your cat the opportunity to express normal behaviors, including stalking and pouncing as if chasing prey. Enrichment can engage your cat's brain (cognitive enrichment) or it can involve their different senses (sensory enrichment). When offering enrichment to your cat, always give them a choice of whether or not to interact with it. For example, if using a new toy or something with scent, put it down and let the cat choose when and if they want to approach it. Forcing something on them could be an unpleasant experience.

TOYS AND PLAYTIME

IF YOU'RE A true cat person, your cat likely has a gazillion toys. Picking out new toys from the pet store can be a lot of fun, but the lovely thing about cats is that they will also play with a crushed-up ball of paper, a stray edamame bean that was accidentally dropped on the floor, or the cardboard tube from a toilet roll. A great interactive toy is simply a hair tie on a piece of string. Cats will get bored of (i.e., habituate to) toys, so it's a good idea to keep them on rotation, and to get new toys from time to time. Providing opportunities for play and predatory behavior is one of the five pillars of a healthy feline environment discussed in chapter 3, and remember that these activities count as enrichment too. If you have more than one cat, you need to make sure each cat's needs for play are met, which may mean individual play sessions in a room away from the other cat(s).

"The world would be a better place for cats if everyone who lived with a cat played with them daily, with interactive toys (e.g., a cat dancer). Many of my behavior consulting clients tell me their cats don't or won't play. But cats are obligate predators and they all (even seniors and cats with disabilities!) have the capacity for interactive play that mimics the hunting experience! I take great pleasure in showing clients how to get their cats stalking, pouncing, or even just mentally engaged with an interactive toy. I think many people don't play with purpose and get frustrated when their cat doesn't respond enthusiastically to a randomly waving feather. But if you move the toy like a prey animal would move, and use all your cat's senses, and remember they are a "stalk-and-rush" hunter, you will have

GREAT SUCCESS (to quote Borat)! So many house cats are leading under-stimulated lives. We give them love, but that isn't enough. They need the benefits of physical exercise (both for staying fit and for reducing stress), and playing with your cat a few minutes a day with a wand toy is a fun bonding experience for you both!"

—MIKEL DELGADO, PhD, certified applied animal behaviorist and cat behavior consultant at Feline Minds and staff scientist at Good Dog

VISUAL ENRICHMENT

MY FIRST CLUE that the hummingbirds have arrived for the summer is not seeing one, but noticing Harley hanging out by the bedroom window where we typically put a hummingbird feeder each year. He spots the birds first, and only then do I see them coming to the window as if to say, "Where's our food?" and I know it's time to put a feeder out. We do see the occasional brave Anna's hummingbird through the winter but it's only in the summer months when we get a lot of them.

Windows with an interesting view provide important visual enrichment for cats. Of course, it's what the cat finds interesting that counts. Some cats seem to spend a lot of time by the window, whereas others don't so much, and it's possible that being unable to reach or interact with things outside the window could cause frustration. In a survey of 577 cats reported in *Journal of Applied Animal Welfare Science*, most caregivers (84%) reported that their cats spent less than five hours a day at the window, with the median time reported as two hours.[2] The most common things for cats to watch from the window were birds, small wildlife, or foliage. Reported less often, but still common, was cats watching other cats, people, vehicles, or insects.

This information about what cats are watching suggests that they would prefer a green outlook that is attractive to birds and other wildlife. If your cats are indoors only, planting your home or patio garden for wildlife such as birds and insects will bring interesting things for your cat to watch through the window. For example, if you feed the birds in winter, position a bird table or birdbath where the cat can watch birds coming to feed and bathe. And place hummingbird feeders within view of a window where cats can see them. In summer, if a window is close to an outside light, you could put the light on at dusk and leave the curtains open for a while. The cat can watch the moths and other insects that are attracted to the light. My cats love this, and Harley will even come to find me if I have forgotten to turn the light on for him. He leaps up at the window trying to catch insects on the other side, so it's a great game that gives him some exercise as well. Another idea is to make sure that cats have a choice of windows, or to let them use a window ledge that is high up, since cats like to be at an elevated position. And if the window can be safely left open, the cat can enjoy the scents that drift in on the breeze.

A window can provide an interesting view.
FIONA KENSHOLE

Catios can be another way of giving your cats an interesting view. Since they are often made of mesh or wire, smells can waft in, even more than through a window. Depending on the layout of your home, it may be possible to convert a porch or balcony into a safely enclosed catio, or to build a catio in the yard with access via a cat flap. Make sure your catio is strong enough to protect your cat from potential predators. Remember to think about providing shade on hot days and warmth in winter. Make use of vertical space by providing walkways and shelves to perch on, as well as hiding places, so that even a small space can make a successful catio. A local woodworker may be able to help, but you can also buy self-assembly catios .

A third option is a cat tunnel and tent designed for outdoor use that will allow your cat some safe space outside. Since these are typically not permanent structures like a catio, it's best if you supervise while your cat uses their tent or tunnel in case they accidentally get caught up in it or tip it over. And remember that outside views and access can have downsides, such as if other pet cats or wildlife pass through the yard and frighten your cat.

Inside your home, television and interactive games (some designed especially for cats) may also provide visual enrichment. It seems likely that videos or TV showing birds or mice will be of most interest. When cats in a shelter were shown different images on television screens for three hours a day, they didn't actually spend that long looking at the screens, and when they did, they seemed to get used to them quite easily.[3] They were least interested in a blank screen (as you might expect) and images of people, and paid more attention to moving images, whether they were animate or inanimate. It seems that images of prey items, or things that move like prey items, would be most likely to work as enrichment. Some of the video game apps available for tablets and cell phones include fish or other objects moving around to tempt cats to try and tap them. Always remember to watch for signs of frustration since the cat is not actually able to catch anything, and end each session by providing an actual physical toy or treat for your cat.

SCENT ENRICHMENT

AS MENTIONED IN chapter 3, cats don't just have an amazing nose for scents; they also have a vomeronasal organ that detects pheromones. That means we can use scent as one way to provide enrichment.

The effects of different scents as enrichment were tested on cats in a shelter run by Cats Protection in the UK.[4] The scents of catnip, lavender, and rabbit were put on a cloth and tried one at a time in cats' cages along with an unscented cloth as one control and no cloth at all as another control. The cloth was added to the cat's cage for three hours a day over five days. Overall, the cats were not terribly interested in the cloths, and their interest did not last long, as they got used to the scents within the three-hour period. The catnip-scented cloth got the most interaction, and many cats played with it by pawing at it, wrestling with it, or rolling on it, which likely reflects a response to the catnip. And when cats had the catnip or rabbit scent, they were less active afterward and spent more time sleeping. The results show that some scents do seem to provide enrichment for cats in shelters, and it's likely that your own cat at home may also like scent enrichment. Remember to always give your cat a choice of whether or not to interact with the scent.

Catnip toys are the most well-known form of scent enrichment. Most cats (but not all) respond to catnip by sniffing it, licking or chewing it, rolling around, rubbing their head or body on it, and seeming to feel euphoric. The active compound in catnip (also known as catmint) is called nepetalactone, and the amount and freshness of it in catnip toys and dried catnip varies. You can, of course, also grow catnip in your garden or on your balcony, where it is a pretty little herb that your cat will probably come and roll on if allowed access. Back in the 1960s, Neil Todd (no relation) showed that this response to catnip is inherited, which is why some cats do not respond that way—they simply did not inherit the response.[5] Now we know that the response does not involve the vomeronasal organ, even though it kind of looks that way, and the parts of the cat's brain that are involved are the amygdala, the hypothalamus, and the olfactory bulb. We don't know why the catnip response has evolved. As for the plant, it seems that catnip evolved nepetalactones to repel insects (and in fact catmint is found in some commercially available insect repellents).[6]

Recently, researchers tested whether nepetalactones may help cats to repel insects.[7] They anesthetized six pairs of lab cats, treated one cat in each pair with nepetalactone, and put their heads in a cage with *Aedes albopictus* mosquitoes. The mosquitoes were much less likely to land on heads that had been treated with nepetalactone. The scientists also found that mosquitoes avoided the cats who had rubbed their head on silver vine, but not other cats who head-rubbed with the vine-rubbing cats, nor cats in a control group. These results show that rubbing on catnip and silver vine is an adaptive behavior that benefits cats, but whether insect repellence is the reason for cats having the catnip/silver vine response has yet to be shown.

Cats respond to some other scents that have not been so well investigated, so scientists tested 100 cats with four substances: catnip, valerian, Tatarian honeysuckle, and silver vine.[8] Valerian is an herb that you can grow in your garden and is found in some cat toys, often combined with catnip. Honeysuckle is a plant, and you can buy the wood in blocks or sticks from some pet stores in Canada, although it seems less available elsewhere. If your cat responds to it, you may need to wash their drool off the honeysuckle from time to time, and if they stop responding, you can always shave a bit off the plant to give it a fresh edge with a stronger scent. Silver vine is known as matatabi in Japan, where it is very popular. You can buy it either as a stick or as a powder, but it's also available as a fruit and fruit galls (where midge larvae have matured). The powder is from these fruit galls and is probably the best type to try with your cat first.

In the study, the cats were presented with the scents in a sock, with an empty sock acting as a control so that when coding the cats' behavior it would not be apparent to the researchers whether or not there was any scent in the sock. Sixty-eight percent of the cats responded to catnip, 80 percent to silver vine, about half to valerian, and about half to honeysuckle. Only six of the cats did not respond to any of the substances. Silver vine contains six chemicals that are similar to nepetalactone, whereas valerian and honeysuckle have one (actinidine).

Dr. Sebastiaan Bol of Cowboy Cat Ranch in Texas, who conducted this research, told me, "This research gave us insight in how many cats in the USA go crazy for catnip and plants that can have a similar effect on cats. Catnip was loved by many, but so was silver vine, a plant that is very popular in Japan, stinky valerian root and the wood of Tatarian honeysuckle. Sadly, about one out of every three cats doesn't like catnip. It's not a choice; it's genetically determined. The good news is that this study demonstrates that most of these cats will love one or more of the other safe plant materials." So it's a nice idea to try these scents with your cat. There are also reports of cats responding to kiwi vine, which is in the same plant genus as silver vine.

"In general, creating a safe, fun, and challenging environment for cats will make them truly happy. Environmental enrichment such as a catio, cat trees, shelves on the wall, hiding places, and food puzzles really make a huge difference for a cat's well-being. But you asked me to name one thing that would make the world better for cats. I am sorry. Plants. Cats love plants (but only the ones that are safe for them). Not only do they really like to eat the grass from oat, rye, wheat, and barley seeds, they also love to lie on or in the grass. Seeding your own grass is easy to do and will give much nicer grass compared to when you buy cat grass in the store.

Other plants that deserve a place in each house with a cat are living catnip plants, silver vine, and Tatarian honeysuckle. Or, to be more specific, wood sticks or powder from the fruit of the silver vine plant, and the wood of Tatarian honeysuckle. They are like catnip, but just a little different. These plants contain chemicals not present in catnip, allowing cats who do not enjoy catnip to have a good time too. Cats who do enjoy catnip may of course still love silver vine and Tatarian honeysuckle as well. The big pieces of honeysuckle wood (stem or branch) look beautiful and will last a lifetime. So, make a difference for your cat today and invest in a huge piece of Tatarian honeysuckle wood, sprinkle some silver vine powder on the carpet, and grow some super fresh cat grass! They'll love you even more!"

—**SEBASTIAAN BOL,** PhD, founder and research scientist at Cowboy Cat Ranch

One of the things indoors-only cats miss out on is the opportunity to smell things outdoors. If you can safely leave a window ajar or a screened window open, this gives them the opportunity to sniff scents wafting in on the breeze. But you can also bring the outdoors in by collecting natural items that are safe for them, such as leaves, sticks, stones, pinecones, and little bits of plants such as mint or parsley (only select plants that are not toxic to cats). Dr. Sarah Ellis calls this a sensory box. She says that normally cats would patrol their outdoor environment to see what has changed, and although they like to keep their indoor environment "familiar to them by constantly scent-marking on it," it doesn't change much. "The indoor environment can be incredibly stagnant," she says, "and there isn't this sort of sensory change. And so we are providing some of that stimulation by bringing some of the outdoors into the home."

Another way to provide scent enrichment is to simply hide food or treats in portions around your home to encourage your cat(s) to forage for it. They may be upset or confused not to receive food in their usual place at first, so begin by just moving it a little way as described by Dr. Sung earlier in this chapter. Once they are used to the idea, you can put it in different places around the house for them to find.

You can also formalize this enrichment into something that's a little more like the kind of beginner nose work people might do with their pet dog. To do this, you will need a few cardboard boxes. Shut your cat out of the room (or put them in another room) and spread three to five cardboard boxes around.

Melina finds a treat hidden in a box. ZAZIE TODD

Put a treat or little piece of tuna, shrimp, or chicken in one of them, and then let your cat into the room. Since cats like boxes, most likely they will start to explore them on their own, but if not, gently encourage your cat to investigate them. As your cat jumps in and out of the boxes, sooner or later they will find the treat. Immediately add an extra treat or two to reward them for finding it. Then do it all again. The first time your cat may be unsure of what is going on, but it will take only a few goes for them to get the hang of it. I've tried nose work with Melina and Harley, and on only the second attempt Melina was already running into the room to look in the boxes. Harley spent some time looking confused. I imagined him thinking, "Why are these boxes here?" But he got the hang of it too, and it became difficult to let only one cat in the room at a time. Just two or three goes is enough to give your cat some enrichment and then you can try again another day.

AUDITORY ENRICHMENT

CATS HAVE GOOD hearing and are able to detect pure tones over a range of 10.3 octaves.[9] Their hearing range extends to both lower and higher frequencies than humans; the high-frequency range is believed to help them hunt by hearing the ultrasonic squeaks of mice. Auditory enrichment could be as simple as a secure open window that lets in the sound of birdsong. If you often listen to music or play a musical instrument, you will probably already have a sense of whether your cat likes music, and if so what kind. Some types of music are designed especially for cats, with the aim of providing enrichment and reducing stress. Examples include Through a Cat's Ear and Music for Cats.

Cats did not pay much attention to Bach or Fauré, but when played Music for Cats they approached the speaker and rubbed their heads on it, showing their approval, according to a study published in *Applied Animal Behaviour Science*.[10] The cat music was at a tempo based on cat sounds such as purring or suckling, and involved plenty of sliding notes similar to the kinds of sounds cats make. Another study, published in *Journal of Feline Medicine and Surgery*, compared silence, classical music, and cat-specific music played to cats before and during a veterinary exam.[11] The cats who heard the cat-specific music had lower Cat Stress Scores than the cats who had classical music or no music; however, there was no difference in physiological measures of stress. If using music with your cat, keep it at a conversational level, not too loud, and avoid any sudden loud noises; also observe your cat to see if they like it or find it stressful.[12]

Interactions with the environment are part of the domain of animal welfare called behavioral interactions, but as discussed in chapter 1, interactions with conspecifics (other cats) and with people are also an important part of this domain. The next two chapters look at these topics.

APPLY THE SCIENCE AT HOME

- Think about different ways you can provide enrichment for your cat that will use their various senses and allow them to engage in normal feline behaviors.

- Always give your cat a choice of whether or not to take part in an enrichment activity; they get to decide if it is actually enriching.

- If your cat isn't interested in the enrichment you provided, think about ways to make it more accessible to them. For example, was a toy too difficult for them? Do you need to practice how to make the toy move like a prey animal? Or did they find it a bit scary? See what you can do to fix the problem.

• Cats will habituate to things, so rotate access to toys and keep providing new enrichment opportunities throughout your cat's life.

8

CATS AND THEIR PEOPLE

· · · · · · · · · · · · · · · · ·

EVERY TIME I come home, Melina walks down the hall to greet me with a chirrup. Her tail is up with a little kink in the end, almost like a wiggly question mark as she walks. She purrs as she rubs against my legs, then moves about a foot away and rolls over, showing her tummy. I know better than to try to touch it. She likes it if I speak nicely to her, and a quick pet on the head is okay, but further attempts at contact are not desired at this time. My tabby cat, Harley, on the other hand, is still where he was when I arrived home—on the heat vent in winter or at the top of the cat tree during the rest of the year. He merely lifts his head a little and watches, waiting for me to go and pet him to say hi, at which point he begins to purr. This has become part of our regular lives. Does this greeting ritual mean they love me?

Love may be a subjective feeling and it's not a very scientific term, but so long as cats were properly socialized as kittens (as described in chapter 2), they like to spend time with people and can form strong bonds with them. This means that, contrary to popular stereotype, many cats do love their people after all. This is probably not a surprise to cat lovers, but it's good to know that scientists have investigated feline love for their owners in several ways.

CATS' ATTACHMENT TO THEIR PEOPLE

IN CHILD PSYCHOLOGY, attachment theory refers to the bond that forms between an infant and their primary caregiver. For infants, this bond is tested using a Strange Situation, a specific set of circumstances that last for a specific length of time. First, the infant spends a short time in a room with their caregiver (typically the mom in this research), then they are joined by a stranger. The caregiver leaves the infant alone with the stranger, then comes back and comforts them, then the infant is left all alone for a short time before being joined again by first the caregiver and then the stranger. The test looks at how the infant responds to their caregiver and the stranger (and to being briefly left alone) in this somewhat peculiar situation. The result tells us what kind of attachment they have to their caregiver. If this test is done with cats, and they respond in a similar way to human infants, it shows they have formed an attachment to their caregiver. As well, the types of attachment can be classified in the same way as for infants.

Many human psychology studies have used the Strange Situation, and they show that most infants (around 60–65%) are securely attached. This means that although they are upset when their caregiver leaves, they are easily consoled and comforted upon their return. Their caregiver is a source of support.

Technically speaking, we say the caregiver is a "secure base" from which they can explore, and a "safe haven" to which they can return if something is stressful. But not all infants are securely attached: some are ambivalent (clingy and anxious), and others are insecurely attached and fairly indifferent to their parents.

You may have already seen signs that you can provide a sense of security to your cat. If your cat likes to lie on your clothes when you aren't there, they are probably enjoying the sense of security they get from the smell of you. Another time you might see that you are a source of security for your cat is when you have visitors to your home. Perhaps when you are there too, the cat is happy to be out in the room and may even visit with the visitors. Your presence is helping the cat to feel safe. But if something stressful happens, the cat may run to be near you again. And if you were in another room, the cat may be less likely to interact with your visitors (unless they know them well) or even come and find you, because they don't feel as secure when you aren't there.

Dr. Kristyn Vitale, assistant professor of animal health and behavior at Unity College, Maine, has investigated cats' relationships with their owners. Studies have already shown that the majority of dogs are securely attached to their owners, just like infants. In a study published in *Current Biology*, Vitale used a shorter version of the Strange Situation to test seventy-nine kittens (aged between 3 and 8 months) and their owners.[1]

"In our case," Vitale says, "what we do is bring the animal into an unfamiliar room, which is our lab, and leave the cat and owner in there for two minutes. And then the owner leaves and the cat's alone for two minutes, and that's a mild stressor. Then the owner comes back, and what we're looking at is the cat's response to that owner's return. Basically, a cat that is securely attached to their owner is going to be able to use their owner as a source of security to go explore out from. And what we found was that the majority of cats, 65 percent of pet cats, were securely attached to their owner. So they used their owner in the same way an infant uses their parents and a dog uses their owner." The remaining 35 percent of the kittens were insecurely attached, of which most were classed as ambivalent (we would call it clingy).

Then Vitale wondered if training would have an effect on attachment style, so half of the kittens took part in a six-week training and socialization class and the other half did not. However, she found that training did not have an effect on attachment. Vitale also tested thirty-eight adult cats (aged 1 year or older) and found that most were securely attached to their owner.

"I think that really what's interesting," says Vitale, "is that they [our findings about cats] are so similar to what we've seen in attachment with humans and attachment with dogs. Not only fitting into those same main attachment styles—so secure, ambivalent which is clinging, or avoidance—but also the proportions of those animals in the population are really similar. Just like in humans, 65 percent of the population is typically secure, and that's exactly the proportion we found in cats and kittens. Some of this might be a biologically relevant behavior that, if you're attached to your parents, they are going to protect you and care for you. So it might actually just be an adaptation of that offspring-parent bond that is kind of being changed to interacting with the owner and living in that same dependency in a human home."

These results are different from the earlier findings of a research team from the University of Lincoln, UK, that tested twenty adult cats in the lab using a version of the Strange Situation test and found no signs of attachment other than cats meowing more when their owner left them compared with when the stranger left.[2] Taking cats to the lab can be stressful for them. So in future research it would seem like a good idea to test cats in their homes. But cats forming attachments to their owners is not the only sign that we are special to our cats.

CATS' ATTENTION AND PERCEPTION

CHOOSING A NAME for a cat can take a lot of thought. Harley was easy; he came with that name and we decided to keep it. But when we brought our tortoiseshell home, we spent a few days without a name, rolling different words around our tongue. She was sweet, so we wanted a sweet name, something mellifluous that would make us think of honey, and so we thought of Melina. She learned her new name very quickly.

Cats are able to recognize their own name, according to research published in *Scientific Reports*.[3] Pet cats were played a set of four words followed by their name, all spoken by the owner. As they listened to the four words, they paid less and less attention, but on hearing their name the cats perked up and oriented to the sound by moving their head and/or ears. Cats who lived with other cats in the home were able to recognize their own name from those of their fellow cats (although when cats in a cat café were tested, they seemed to respond to their own name and those of the other cats). We also know that cats can tell the difference between their owner's voice and that of a stranger.[4] So the researchers then tested whether cats still recognized their name when spoken by a stranger, and they did. Cats may learn to associate their name with positive things about to happen if their guardian calls them and then pets them or gives them a treat, or with something stressful such as their guardian calling them prior to taking them to the vet.

A study of 6,000 interactions between pet cats and their female owners in the home had some interesting findings, as reported in *Behavioural Processes*.[5] Surprisingly, the more successful the person was at trying to initiate contact with their cat, the shorter the interaction was. However, the more the cat was responsible for initiating interactions, the longer those interactions were. These results show it's best for interactions to be on the cat's terms. Earlier research between lab cats and volunteers who were brought in to meet them and first instructed to spend five minutes reading a magazine and ignoring the cat, then to spend five minutes trying to interact with the cat, showed that cats notice when people are paying attention to them.[6] When people tried to interact with the cat, there were fewer vocalizations (suggesting these had been attempts to get attention), and more play and head-rubbing by the cat.

Another study, also published in *Behavioural Processes*, found that both pet cats and shelter cats choose to spend more time near a stranger who is attentive rather than inattentive.[7] The person playing the role of an attentive stranger would call out to the cat and pet them as long as they were in reach, whereas the person being an inattentive stranger would merely pet them twice without looking at or speaking to them. The cat soon chose to spend time with the stranger who was offering lots of fuss. This shows that cats know when the person is paying attention to them, and they respond to it. The person's attention (or lack of it) did not affect how much the cats meowed, however. And the shelter cats spent more time than pet cats did with the inattentive stranger, which could be because they are deprived of human contact. In a follow-up study with pet cats only, they did not spend more time with their owner than with a stranger, and did not meow any more or less in response to their owner. But this does not mean the owner is not special to them, just that a well-socialized cat is also friendly to strangers.

Social interaction with people is important to cats (so long as they have been well socialized). When researchers gave cats a choice between four different stimuli, social contact with a person was preferred by half (50%) of the cats, according to a study in *Behavioural Processes*.[8] But 37 percent of the cats picked food, with 11 percent choosing a toy and just 2 percent choosing something scented. Cats purr more and stretch more when reunited with their guardian after four hours apart compared with an absence of just thirty minutes (incidentally, the person also spoke to their cat more after a longer gap).[9] And studies also show that cats are sensitive to our moods and emotions, preferring a happy person to an angry one.[10] When the owner shows positive emotions through their face and posture, compared with when they are angry and frowning, pet cats spend more time near them and engage in more positive behaviors such as purring and rubbing against them, according to research published in *Animal Cognition*.[11] But perhaps surprisingly, cats spent equal amounts of time with a stranger whether they were showing positive or negative emotions. Although the cats did distinguish between happy and angry voices from their owner, the difference was not as notable as for posture. These results suggest that cats learn how to recognize their owner's emotions. Perhaps this is because they sometimes have consequences for the cat.

Back in 1998, when Ádám Miklósi (of the Family Dog Project in Hungary) and Brian Hare (now of Duke University, North Carolina) independently published papers showing that dogs can follow people's pointing gestures, it made ripples across the scientific establishment. This finding is widely credited with being one of the things that kick-started scientific interest in dogs. Then in

2005, Miklósi and colleagues published a paper in *Journal of Comparative Psychology* that showed 70 percent of cats could follow when an experimenter pointed to one of two bowls that contained a treat, and they would go to that bowl and get the treat.[12] Cats did best when the pointing finger was relatively close to the bowl, and not quite as well when it was farther away, but still, most of them could follow the point. As well, in a further experiment when the food was made inaccessible, they found that 42 percent of the cats would gaze towards a person as if asking for help in accessing the food. Although that's not a majority, it's still pretty impressive that so many cats can use this means of communication, especially when you consider that we are another species.

In a further study, scientists were able to show that cats can also use human gaze to identify which of two bowls contain food. You can try this with your cat at home. Set out two bowls, one of which contains a treat, and then point at it to see if your cat will go to that bowl. You can then try pointing from farther away to see if your cat still goes to the right bowl. And then later (or another day), you can see if they can also follow your gaze to pick the bowl with the treat. This could be a fun game to test your cat's cognitive abilities. Making the treat inaccessible, like in one of Miklósi's experiments, will probably not be as much fun for your cat! And don't worry if your cat doesn't successfully follow the point; remember that some of the cats in Miklósi's study couldn't either.

Taken together, this body of research shows that cats pay attention to us, notice our gestures, and listen to what we say—all signs of how social they are towards people. They know their name and can recognize their owner's voice. And it turns out there's something interesting in how cats communicate with us too.

MEOWS AND CHIRRUPS

I CAN HAVE quite a long conversation with Melina. "Meow," she says, and I meow back. "Meow," she says, and I think she agrees with me so I confirm, "Meow." Then she meows again. Often they are short meows, but sometimes they are longer and more drawn out like "meoooow," and when I reply with the same, she seems to match what I just said. She is living up to the reputation of tortoiseshell cats being chatty, but at the same time I think she is also showing signs of communicative behaviors. Harley is not so chatty, and his meows serve a different purpose: asking me to feed him, or brush him, or feed him again.

Either way, there is something special about a pet cat's meow: it's designed to communicate with humans. The chirrup type of sound that is used to greet people is used by mother cats to call to their kittens, and kittens will meow when separated from their mother. Adult cats only very rarely meow at each other, and the meow seems to be specific to a particular cat and their human. According to a study published in *Anthrozoös*, people are very poor at identifying what the meows of an unfamiliar cat mean.[13] The scientists tested meows recorded in four different contexts: when the owner was preparing food, when they were withholding food, when the cat was asking for attention, and when they were trying to deal with a barrier that separated them from their owner. With an unfamiliar cat, people could not match the meows to the context. When it was their own cat, however, people performed better than chance, showing we are tuned in to our own cat's meows. What this suggests is that cats don't have a general set of meow sounds that they use in different contexts; rather, each cat has their own unique set of meows, which their person often learns to identify.

Perhaps the purr is the nicest form of feline communication. A cat's purr is a beautiful, rumbling sound. Kittens purr when nursing, as does their mom while they suckle. Purring is very low-pitched, with an average frequency of 27 hertz. As kittens are born deaf and blind, it's thought they use the vibrations from the purr to help them find their mother, and purr back at her so she knows they are content. Cats will also purr in interactions with humans, often to show contentment like when they are sitting on your lap being petted. They also purr in interactions with other cats, such as when resting with or rubbing on another cat they are friends with. And it's important to know that cats will sometimes purr when in pain.

A study published in *Current Biology* shows that cats also have a solicitation purr—a purr that's used to ask for something such as food at mealtimes.[14] The acoustic qualities of this purr are different, including some frequencies (300–600 hertz) similar to those in a baby's cry. As a result, people rate the solicitation purr as less pleasant and more urgent than a regular purr. When played two different purrs from the same cat, most people are able to distinguish between the regular purr and the solicitation purr. The scientists also showed that when they technically altered the purr to remove the bits of the signal at the higher frequency, people no longer found it so urgent (although they didn't think it was more pleasant).

Cats' vocalizations are therefore designed for us, or to communicate specific needs to us. The idea that each cat has their own unique meows that only their owner can understand is just part of the idiosyncratic relationship everyone has with their cat. And that purr often accompanies something that brings pleasure to both cat and human—petting.

HOW TO PET CATS

ONE OF THE lovely things about having a cat is being able to pet them. Feeling the softness of their fur and listening to a full-throated purr brings a lot of satisfaction. But it's up to us to know if they want to be petted in that moment, and where they prefer it. If you've made the mistake of trying to pet a cat on the tummy, you will probably have learned the hard way that when they stretch out and show their tummy, they are not asking you to pet it. Although every cat is an individual, most cats have preferences about where they would like (or not like) to be touched. And it turns out science can help.

Researchers tested where cats prefer to be petted, and in what order, in two different experiments published in *Applied Animal Behaviour Science*.[15] The results show that cats have definite preferences. It is thought that animals prefer petting from humans to be similar to the ways they show affection to members of their own species. We know that friendly feline behavior involves parts of the body with many scent glands: around the lips, chin, and cheek (the perioral gland); between the eyes and ears (the temporal gland); and near the base of the tail (the caudal gland). When cats rub against each other in these areas, they are transferring scent and building up a group scent (see chapter 9). So you might guess these would be the areas cats prefer to be petted by people. There could also be an order effect, because when cats rub on each other, there's a definite direction: rubbing their heads together first, and only sometimes progressing to intertwining tails.

The researchers tested thirty-four cats in their own homes. After some time to get used to the experimenter and the video recorder, cats were tested on two days, one on which the owner petted them, and one on which the experimenter petted them. As well as the three scent gland areas, they tested five other parts of the body (top of the head, back of the neck, top of the back, middle of the back, and chest and throat area). The experiment was standardized: the order of body parts to be petted was random, petting was done with two fingers and lasted for fifteen seconds in each area. However, cats were free to walk away at any time. And, being cats, many did. Only sixteen of the cats were stroked in all eight areas by both people.

The researchers analyzed the videos to see how many times friendly behaviors occurred, such as giving a slow-blink, licking the person, rubbing their head against the person, grooming, kneading, and holding the tail straight up or up with a curl on the end. And they also counted how many times negative behaviors occurred, such as swishing or flicking the tail, moving the head away from the person, licking the lips, biting, or cuffing the person with a paw. There were no differences in positive behaviors, but when being petted by the experimenter, cats showed more negative behaviors when petted near the tail. In other words, they didn't like this so much. As well, the cats seemed to prefer being petted by the experimenter more than by the owner. This could be because the experimenter was new and therefore interesting to them, or it could be that they were not used to the different way in which the owner petted them (remember, the owner had to use two fingers to pet the cat in order to standardize the experiment).

In a second experiment with twenty different cats, owners petted their cat in a set order, either from the top of the head and along the back to the tail, or vice versa. This time they could use their whole hand or just one or two fingers, however they would normally pet the cat. And this time, only three cats moved away. These videos also showed that cats did not like being petted near the tail, regardless of the order. But they did not seem to mind which order they were petted in. The lack of an order effect suggests that being petted is more like allo-grooming (grooming another cat) than allo-rubbing (rubbing the body against another cat), though more research is needed.

So what does this mean for the human-feline relationship? The scientists say owners should avoid petting near the tail. Instead, they should pet the face, especially in the areas where the scent glands are. It could also be that cats like interactions with their owner to be on their own terms and prefer to initiate the interactions themselves. It is important to give cats a choice when we pet them, and to learn what your particular cat likes (maybe your cat does like being petted near the tail or on the tummy). It's also possible that some cats learn negative associations to their owners; for example, if the owner scolds them. This is one good reason why we should train cats with food or brushing as a reward, and avoid the use of punishment (see chapter 5).

So if you want to pet your cat, give them a choice and find out if they would like it or not. You can put out your hand, fist, or a finger and see if they choose to approach. It may also help to get down to their level, especially if the cat is timid. If they approach and sniff but then move away, they don't want you to pet them. But quite likely they will approach, sniff, and then begin to lean in or rub their head on you. This means you can begin to pet them with your hand. Remember, they generally prefer to be petted on the head or face where the scent glands are; some cats like to be petted under the chin, and will move their head just so to ensure you get the right spot. Melina is such a cat.

Signs that the cat enjoys the petting include a purr, half-closed eyes (or even completely closed), and the cat leaning in. If unsure, you can always do a consent test: stop petting and see what happens. If they walk away, the session is over. But if they want more, they will make their feelings known by rubbing their head on you or pawing at you. And remember to always watch the body language. Since most cats prefer petting sessions to be short, you need to stop before they get too aroused or fed up with you. I always keep an eye on the tail, because a twitching tail can be a sign that the cat may have had enough, especially if it's a widely swishing tail. Other signs to look for include dilated pupils, twitching skin, pushing your hand away with a paw, getting the claws out, trying to scratch you, or looking at your hand and fixating on it. If you keep petting too long, you risk a bite or scratch as the cat makes their feelings known. Remember that hugs and kisses are quite intense forms of interaction and too full-on for most cats.

With children, take care to ensure interactions are happy for both child and pet. Young children are still learning motor control and may be accidentally too rough, so you should guide their hands to help them. Children also tend to go for hugs and kisses, which as just mentioned are usually too intense. Teach them to give the cat a choice and a chance to walk away, and ensure the cat has access to spaces that are out of reach of the child. Younger cats (up to 6 years) are more likely to be affectionate to children, and old cats less so; as well, just because a cat is friendly to adults does not mean they will also be friendly to children.[16] It seems that all family members (human and nonhuman) are important in predicting whether a cat will be affectionate towards children, and that cats do better with children when there is more than one cat in the home. As well, cats from breeders and shelters seem to get on better with children than cats acquired via newspaper ads, although few cats are aggressive.

"Sadly the social behavior of cats, and especially their interactions with people, is very misunderstood. Most cats typically want high-frequency but lower-intensity interactions whereas many people want fewer interactions but for a longer period of time. This mismatch can lead to defensive aggression in cats with some being labeled as grumpy or spiteful. Having more realistic expectations around the interactions that cats appreciate—frequent but short—will avoid unnecessary stress, fear and worry and will help strengthen the bond between cat and owner."

—**DR. SAM GAINES,** head, Companion Animal Science and Policy, RSPCA

CATS' SOCIABILITY TO PEOPLE

CATS HAVE REALLY not been domesticated that long when you think about it in evolutionary terms, so it's interesting to consider how they have learned to be so sociable with humans. In a review of the literature in *Journal of Veterinary Behavior*, biologist Prof. John Bradshaw suggests three ways in which cats have developed to communicate with their owners.[17] The first involves behaviors the cat already knows as a species and that have become associated with positive outcomes, such as jumping up on the owner. The second category is signals that kittens use to communicate with their mother and that are then used with their people. This includes the kneading of paws that you often see when a cat is sitting, purring on your lap; this is similar to the kneading of paws that kittens do when suckling from their mom. The meow would also fall into this category, given that adult cats do not meow at each other. The third category is signals that have their origin in communication with other cats, such as rubbing the head on people's hands or legs, which is similar to the way they would rub their head on cats they are friendly with.

Despite the popular stereotype of cats as aloof, it turns out they show affection for their people in numerous ways. I asked Dr. Kristyn Vitale how people can tell if their cat likes them. "One thing we always find in our research is that there's a lot of individual variation with cats," she said, "and so cats might show you that they like you in different ways. If your cat is a really sociable cat that's really bold, it might show you it loves you by sleeping on you every day or hanging out with you, versus other cats who are maybe more solitary but still might like you, but they might show it in a different way such as sitting across from you or away from you but purring and meowing at you."

You are probably used to seeing a lovely, slow eyeblink from your cat, in which a series of slow half-blinks is followed by the eye closing mostly or fully in a slow-blink. This is a sign of a relaxed cat, and a lot of people like to slow-blink back to their cat. This is a good idea according to scientists who tested cats in their own homes.[18] First, they taught the guardians how to do the slow-blink, namely to slowly close their eyes while making sure not to lower their eyebrows or wrinkle their nose but to move their cheeks upwards. This is similar to the movement the cat makes. Then, they tested what happened when guardians slow-blinked at their cat. The cats narrowed their eyes in response to slow-blinks much more compared with when the owner was around but not interacting with them, and they did not do so in response to normal blinks.

In a second study, they found that cats were more likely to approach an experimenter who slow-blinked at them compared with a neutral condition. This is especially nice because it shows cats responding to a slow-blink from someone they don't know. I am already used to doing a slow-blink when I meet cats, and probably so are many others who work or volunteer with cats, so it's nice to see this research that shows cats seem to like it. Next time you look at your cat, give them a slow-blink and see what happens.

The research on cats' attachment to and sociability with people has several implications for those of us who live with cats. For one thing, we may be a "safe haven" for our cats when something stressful happens, so our presence (and comforting behaviors) may help them to cope with stress. But this means we also need to work to maintain that role; people who resort to punishments such as telling their cat off or spraying them with water will be damaging that relationship. And since cats enjoy affection from us, we should pay attention to how they like it, and aim for frequent, short interactions with them.

When it comes to cats' social relations, we also have to think about the nonhuman animals—especially other cats and dogs—who share our home, which is the topic of the next chapter.

APPLY THE SCIENCE AT HOME

- If your cat doesn't seem to know their name, teach them by saying it and then following it with something nice like a treat or some petting.

- Aim to have multiple short interactions with your cat each day, as that is what they prefer. If you can make at least some of these interactions predictable, that's even better (e.g., when you return home, you always wait for your cat to approach you and then pet them; the cat jumps on the bed every morning and you pet them before getting up).

- Pay attention to your cat's body language and learn how they prefer to be petted. Every cat is an individual, and it's up to you to know what your cat likes. That said, remember that most cats prefer to be petted on or around the head, and for most cats the belly and tail are out of bounds.

- Give your cat a choice in whether or not to be petted, and explain this to visitors if necessary.

- Let your cat initiate interactions with you.

- Remember that some cats are lap cats, others prefer to sit next to you on the sofa, and others are just content to be in the same room. Whichever kind your cat is, that's okay. If you want to try and train your cat to sit on your lap, reward them with petting or a treat for approaching or getting on your lap, but don't ever force them to be there. Ultimately, we have to appreciate cats for who they are.

9

THE
SOCIAL CAT

....

HARLEY AND MELINA did not arrive at our house together. Harley came first, and having seen him being sociable with other cats in a communal room at the shelter, we felt sure he would like a feline friend to keep him company. But we knew we had to choose carefully, as cats don't always get along with each other and some cats prefer to be the only one in the home. At the shelter, there was a tortoiseshell cat called Lisa-Marie who stuck her paws through the cage at us when we were in another room meeting a different cat. When we went into her room—because we wanted to meet all the cats before making a choice—she rushed over to rub against our legs and purred loudly. We spent a little while with her and she charmed us right away. We went back into the corridor to consider the cats we had met. Another couple stopped and crouched down by the glass window to Lisa-Marie's room, trying to get her attention, and she ignored them completely. But when we approached the glass to get another look at her, she came up and put her paw to the window. Our hearts melted.

The staff said they thought Lisa-Marie would likely get along with another cat, so we chose her, feeling that she had been the one to choose us. She had been in the care of the shelter for many months, having arrived as a pregnant stray cat, delivered her kittens in a foster home, and developed complications from the spay surgery that followed. And then she got an upper respiratory tract infection. After all this, we knew this sweet tortoiseshell—whom we renamed Melina—might take a little while to settle in.

When introducing two cats, make the process very gradual and only allow them to see each other once they are already used to each other's smell. Because of the layout of our house, keeping Harley and Melina apart for long wasn't going to be easy, so we went through an accelerated version of this approach. Luckily, both cats were happy to accept each other's smell and seemed okay when we introduced them to each other. Of course, it helped that we had already set the house up so that the two cats did not need to compete over resources. Before long, I found myself walking into a room only to see Harley and Melina engaged in some long, mutual licking process and springing apart like teenagers caught French kissing. Over the years, there have been occasional moments of tension, especially one time after Melina had to keep going to the vet for a painful issue that made her cranky, and also made her smell of the vet. But I'm pleased to see them get on well, given that they share the same house and, as indoors-only cats, cannot choose to live apart from each other.

As mentioned in chapter 1, around a fifth of cats live in a home with another cat they don't necessarily get on with. It's not always easy to tell if two cats will get along, and there are no guarantees, as every cat is an individual.

CATS AND OTHER CATS

THE SOCIAL STRUCTURE of cats is fascinating but not well understood. Cats hunt on their own, and a mouse—a typical meal for a cat to catch—is not big enough to share. Cats can manage quite happily on their own and many prefer it. However, cats can also live in social groups, and this is especially the case for groups of female cats. This diversity in social relationships means that cats can vary quite a lot.

When the environment has a lot of food and can support many cats, cats will live in large groups, called colonies, according to a review in *Journal of Feline Medicine and Surgery*.[1] With a more limited amount of food, the colonies will be smaller, and if food sources are very spread out, then the environment will only support a solitary cat. Cat colonies are matrilineal, with friendly relationships amongst female cats providing the structure to the group. Members of a colony know whether another cat is part of their group or not and will see off cats from outside. It is possible for new cats to join, but doing so is a gradual process involving many interactions over a long period of time.

Even within a colony, relationships are not all the same; cats have preferences over who they like to hang out with, and it is these preferred associates who are often found within 1 meter (39 inches) of each other. These cats will greet each other with a nose-touch, will allo-groom (groom each other) and allo-rub (rub their bodies alongside each other), and intertwine their tails. They also approach each other with the tail up, showing friendly intentions. Allo-grooming and allo-rubbing transfer scent between the cats. It is thought (though not known for sure) to maintain a "colony odor" that helps define who is in the group and who is not.

Female cats often cooperate with each other in caring for kittens, even when they are not related to each other. This care extends to unspayed females, known as queens, helping another cat through birth, including licking the other cat's kittens to clean them and eating the placenta. Queens will also help to take care of another cat's kittens, nurse kittens from other cats (allo-nursing), and give them prey, as well as take food to a mother cat who is nursing her kittens. This communal care is advantageous. If kittens have to be moved suddenly, other cats are around to help. And kittens from this kind of care leave the nest ten days earlier than kittens brought up by a mother cat on her own. Unneutered male cats, known as toms, can also be preferred associates of each other and do not necessarily fight, even in the presence of a female cat in estrus whom they may both mate with.

Both the tail-up signal and allo-grooming seem to be important in maintaining social cohesion within a group of cats, so if you see your own cats doing these, they're a good sign. It seems likely these signs developed from the mother-kitten relationship and then, so long as there is enough food around, they continue as ways of showing affiliation.[2]

The social lives of feral cats tell us a lot about what cats in homes would like. Pet cats can form social groups despite not being related to each other, but they need to have enough resources and especially must not have to compete for food (see chapter 3). As well, to introduce a new cat requires many short interactions (see page 151), and cats do not want to have contact with other cats who are not part of their social group.

Although it is not possible to predict whether one cat will get on with another, some clues can help us assess if it is likely. If you're choosing an adult rescue cat, ask for information about the cat's previous life. And if you're getting a kitten to go along with

Cats who are part of the same social group will share. Here, Rupert has brought Grace a mouse. (The mouse was rescued.) *FIONA KENSHOLE*

Cats in the same social group will choose to spend time near each other and even touching. *JEAN BALLARD*

an adult cat you've owned since kittenhood, ask yourself some honest questions about your cat's experiences. Because early life experiences are so important, it helps if, as a kitten, the cat had positive experiences with other kittens and saw their mother being friendly and sociable too (see chapter 2). A cat who was adopted as a kitten into a home where they are the only cat will have missed opportunities to learn about social interactions with other cats during late kittenhood and the early juvenile period. As a result, they may not have great social skills and are not so likely to be friendly with another cat later on in life. Spayed female cats are the most likely to be aggressive to other cats.[3] If you know that you will want multiple cats, it often makes sense to get a group of two or three related kittens/cats all at once. Another factor to take into account is the personality of the cat you already have and that of the cat you are thinking of bringing into your home. For example, a bold cat and a shy cat may not get on if the bold cat bullies the shy one.

"If you are bringing an adult cat into an existing household with one or more adult cats, be prepared that it may not be an overnight success," says Dr. Beth Strickler. "It may be a process. I encourage people to think of it like finding a compatible room-mate or a compatible spouse. They don't always get along." She recommends bringing in the new cat as a foster or for a trial period if possible, so that the cat can go back if it turns out one of the cats doesn't like the other cat (or doesn't like any other cat). She says you need to be prepared that "you're not going to probably bring this cat home and your cats are going to be best friends tomorrow."

According to a review published in *Applied Animal Behaviour Science*, cats will have better welfare in homes with more than one cat if you ensure they all have access to enrichment.[4] Although cats may share large resources (e.g., multiple cats lounging on your bed during the daytime or sharing a large cat tree), some resources are small and may be monopolized by individual cats. Each cat tends to have particular areas where they prefer to spend time, and although cats are pretty good at time-sharing, it is bet-ter if they all have access to places they like to hang out. One study mentioned in the review found that cats will play more when they have more space available to them. Of course, no one expects you to move to a bigger apartment or house for this purpose! You may be able to increase space simply by ensuring doors to rooms are left open and not blocking off any rooms from the cats.

To introduce a new cat, you need to make them feel like they are already friends before they actually meet each other. You can do this by letting them have low-intensity contact, one sense at a time. To start, for example, take some bedding that smells of the new cat and show it to the existing cat (and vice versa), and pair it with treats to make a positive impression. Only once the cat is content with the new smell should you progress to another stage, perhaps allowing a brief distant sighting or sound of the other cat while they are distracted by a toy. Make the process very slow and gradual, and only move forwards when you know both cats are happy. It may be tempting to take something that smells of the old cat and rub it on the new one (or vice versa), but this approach does not give the cat a choice. Choice and control are important, as well as keeping the initial interactions brief.

If your cat prefers to be an only cat, it is best to have only one cat. And it's important to know that cats in the same home may have a lot of tension, even if it is not immediately apparent to the owner. Cats who get along with each other will choose to spend time close to each other, and may even cuddle up together and groom each other. Cats who are friends like this will often share resources. But cats within the same household may prefer to be separate. Sometimes if they are sharing a bed or settee, it simply reflects the lack of alternate places to lounge, rather than a friendly relationship. Dr. Lauren Finka told me, "A lot of people will just focus on physical indications, so they'll put a lot of weight on whether the cats are actually fighting. Are they hissing at each other? Are they chasing each other away? And they're really important signs to focus on, but of course cats can be very subtle and they can bully each other from a distance. Actually what you see is a lot of cats just passively avoiding each other, or one cat will maybe stare at another cat from across the room, or just place itself in between another cat and an important resource. And if you look out for them they're pretty obvious, but if you're only looking out for the physical fighting and hostility, you're probably going to miss all these other signs which are really, really important as well."

If you do see these subtle signs, all is not lost. Start by taking a closer look at the environment and see if you can find ways to reduce stress. Make sure they are not competing over resources, that the cat litter is scooped often, and that each cat has their own quiet space in which to eat meals. You may also decide to try

Feliway Multicat (also called Feliway Friends), a plug-in diffuser that uses synthetic cat-appeasing pheromones to promote feline friendliness.

One study, published in *Journal of Feline Medicine and Surgery*, put this pheromone to the test in a double-blind, randomized, controlled trial.[5] Because Feliway Multicat is a plug-in diffuser, the placebo was designed as a plug-in diffuser that looked identical. People with two to five cats in the home and who reported aggression between their cats took part in the study; seventeen people were assigned to the Feliway Multicat group and twenty-five to the placebo. People in both groups attended a ninety-minute training session with a veterinary behaviorist that provided information on cat behavior—including how to recognize aggression and how to tell the difference between play and aggression, information on counter-conditioning, and instructions not to use punishment (e.g., spray bottles) or startle their cats. Participants drew a plan of their house, and the scientists decided the best place to plug in their diffuser (either Feliway Multicat or placebo). The diffusers were used for twenty-eight days during which cat owners filled out a daily diary and a weekly questionnaire about their cats' behavior.

Rates of reported aggression dropped even before the start of the treatment, likely due to the education session. Aggression also decreased in both groups during the twenty-eight days using the plug-ins, but it decreased significantly more in the group using Feliway Multicat. During a two-week follow-up, aggressive behaviors remained low in the group that had used Feliway Multicat but began to go up in the placebo group. There was no difference between the two groups in terms of affiliative behaviors such as nose-touching, sleeping in the same room, or licking the head or neck of another cat. At the end of the study, 84 percent of the people who had used Feliway Multicat said their cats were getting along better compared with 64 percent of those who had the placebo. These results are based on owner observations, and even though they received training, it is possible people missed some behavioral signs.

Owners who use synthetic pheromones without learning more about cat behavior may not see the same results as the study. If you decide to try Feliway Multicat, place the diffusers where your cats spend a lot of time. Be sure to read the instructions on where to place them (not underneath a shelf, for example) and how often to replace them so that they aren't a fire risk.

FELINE SOCIAL BEHAVIOR AND PLAY

NOTHING IS SWEETER than seeing cats who are housemates cuddle up together. As mentioned, cats in the same social group will choose to spend time close together and may engage in head-rubbing, body-rubbing, tail-wrapping, and allo-grooming. Another important feline social signal is the tail-up signal. This is when a cat approaches another cat with their tail held vertically, perhaps with a small kink at the top where it bends over. In the domestic cat this is an affiliative signal, as shown by Dr. Charlotte Cameron-Beaumont in a PhD thesis for the University of Southampton, UK.[6]

First, from a database of interactions that were observed between feral cats, she analyzed all the interactions that started with the tail-up signal. If the other cat responded with tail up, then the cats were likely to rub on each other. Some cats did not respond with tail up, and on those occasions, the cats did not tend to rub. Then Dr. Cameron-Beaumont did an experiment involving pet cats in their own homes: she fixed the silhouette of a cat with tail up or with a neutral tail to a wall and watched to see how the cats responded. Of course, she could do this test only once with each silhouette (on different days) because the cats quickly realized that it was not a real cat. But when looking at those first sightings, she found that cats approached the tail-up silhouette faster than the neutral one, were more likely to have their own tail high when they first saw it and when approaching it, and were less likely to wave their tail around, compared with the neutral silhouette. This research suggests that tail up is a friendly signal.

Observations of interactions between feral cats in a cat colony in Rome, published in *Behavioural Processes*, confirm this.[7] In this study, tail up was followed by rubbing or nose-sniffing only 23 percent of the time, but nonetheless it was seen as an affiliative behavior. Female cats were most likely to take the initiative in tail up and rubbing, and male cats were more likely to initiate nose-sniffing.

Play for cats is largely based on predatory behavior, in which they treat an object as if it were a mouse, insect, or other potentially edible item. Cats do not seem to mind if the object is being manipulated by a person, such as when they play with a wand toy. Yet compared with other animals, cats will play more when hungry, rather than less, according to *Journal of Feline Medicine and Surgery*.[8] This means that if you want to engage your cat in play, they are especially likely to be receptive to the idea just before they get fed. As well, when cats are hungry, they will play with bigger toys than they normally would, just as hunger would lead them to be bolder in the kinds of prey animals they chase. When cats play with toys, they play with them exactly as they would if the toy were real prey. For example, they pounce on a toy mouse as they would a real mouse, or carry the toy away in the same way they would carry away a captured mouse.

Types of play in cats

TYPE OF PLAY	WHAT IT INVOLVES	AGES
Social play	Play with other cats (littermates and mother)	• Begins at 2–3 weeks • Peaks at 9–14 weeks • Decreases between 12–16 weeks as the kitten becomes more interested in prey
Object play	• Play with objects • Includes aspects of predatory behavior such as pouncing, biting, batting • Often occurs in sequences, and often begins by poking or batting the object	• Interest in moving objects begins at 4 weeks • Play peaks at 18–21 weeks
Locomotor play	Play involving movement and exploring the environment; e.g., climbing things	Begins at 5 weeks
Predatory play	• Play with prey • Develops based on the mother's behavior; the mother brings prey from the beginning of weaning and demonstrates what to do. Kittens with a mother are quicker to bite and carry mice, and kill more mice, than kittens separated from the mother	Increases from 6 weeks

Sources: Delgado and Hecht (2019); Bradshaw, Casey, and Brown (2012)[9]

Social play between cats may peak during kittenhood, but adult cats will continue to play together. A common question for cat guardians is whether their cats are playing or fighting, and some signals will help you know. Kittens have a "play face," with the mouth half-open, that they often make at the start of play. Another play signal is the side step, a little sideways step they take with an arched back that is often followed by a side step from the other kitten. More commonly, play will begin with a pounce, stand-up (standing on the hind legs), or belly-up (showing the belly). All these signs can be seen in adult cats too. Play typically remains as play, but much activity and wrestling is involved, although it can sometimes tip over into a fight. The claws will remain sheathed during play. As well, expect to see frequent role reversals, as kittens alternate who does stand-up and who does belly-up. Play often comes to an end with a chase sequence or a vertical leap.

These play signs are different from real fighting, where there may be caterwauling, hissing, and spitting; the cat's ears will be pinned back; and their claws will be out. Be aware of subtler signs of aggression, which include a stare, piloerection (hair standing on end or a bushy tail), and the cat's ears turned back or flattened against the head. Signs such as pouncing on the other cat, batting at them, biting them, or jumping on top of them are more obvious. Sometimes fights will appear to happen in slow motion as one cat tries to slowly back away to avoid confrontation.

In the unfortunate event that two cats are having a real fight, try and distract them or slide something (like a piece of cardboard) between them. Attempting to separate them risks you getting a redirected bite. Real cat fights can cause a lot of damage. If your cat is injured in a fight with another cat, take them to the vet to get any wounds assessed and cleaned and to determine if antibiotics are required due to infection. Similarly, if you are bitten, seek medical attention, as cat bites can easily become infected. You may need a tetanus booster, and depending on the circumstances and location, you may also need a doctor to assess the risk of rabies.

"The one thing that would make the world better for cats, I think, might be for us to have a better understanding of how they see the world. Unlike us, and unlike dogs, they aren't naturally gregarious as a species. This means that whilst some cats like some other cats, for the most part (there are always exceptions...) cats don't want, like, or need lots of cat buddies. So when we keep them as pets, one cat in a household is fine, and certainly doesn't need to have extra friends! Also, cats are nature's control freaks and need some self-determination.

One of the most common problems I see is when people forget that feral cats are truly wild animals and, meaning well, try to tame them. This means the cat gets guaranteed food, shelter, and veterinary care but at the cost of being able to make its own choices about proximity to humans and other animals. For most feral cats, this probably isn't overall a beneficial trade-off, any more than it would be for a weasel, badger, or other wild animal. So, people respecting and understanding cats' fundamental needs to generally behave as solitary psychopaths would, I think, make a better world for cats."

—DR. JENNY STAVISKY, assistant professor, Shelter Medicine, School of Veterinary Medicine and Science, University of Nottingham

CATS, DOGS, AND OTHER PETS

MY LATE DOG Bodger loved both Harley and Melina and was always happy to see them. He especially liked Melina, and whenever she appeared he would walk up to her with his tail wagging and they would sniff noses (an image of this, drawn by artist Lili Chin, is the logo for my blog, *Companion Animal Psychology*). They did not always get on perfectly, of course. As a herding dog, Bodger would sometimes herd Melina and Harley out of one room and into another, which was comical to watch but I think the cats sometimes found trying. And there was a phase when Melina would creep up on Bodger while he was sleeping and sniff his head, which made him leap in the air out of fright. I took to saying "Melina incoming" to Bodger when I saw her coming, and he started lifting his head and looking for her in response. In turn, Melina stopped doing this, leaving me to wonder whether, having discovered that it would cause such a strong reaction, she was trying to recreate it. But overall, although they never cuddled up together, they greeted each other happily many times a day. We trusted Bodger with the cats, but never left them unattended together when we were out of the house.

Some dogs are a risk to cats. A high prey drive means that some dogs (such as many sight hounds) will think of cats as food and try to catch them, or chase and catch them if they run. Such dogs can and sometimes will kill a cat. So if you want to have a cat and dog in the same home, be careful (and realistic) in your choice of dog. Don't take any chances if the dog is a potential risk to the cat. But dogs and cats can get along well, and a couple of studies look at this relationship.

A study published in *Applied Animal Behaviour Science* provided some good news about cats and dogs who live in the same home.[10]

The scientists distributed a questionnaire to pet owners who had both cats and dogs, and also spent time in the home observing how the cat and dog interacted when in the same room. In approximately 66 percent of the cases, the cats and dogs showed amicable behaviors towards the other animal. They were aggressive in less than 10 percent of the cases, and the remainder were indifferent. Dogs were more likely to be friendly to the cat if the cat was adopted first. They were more likely to have a friendly relationship if introduced at a young age, which for cats was less than 6 months and for the dogs was less than 1 year old.

One very nice finding from this study was that the cats and dogs often seemed to understand each other's communication, even though they use different signals. For example, a wagging tail is a sign of friendship from a dog, but of nervousness or impending aggression from a cat. But the cats and dogs seemed to be able to read each other's body language. The dogs had even learned a cat-friendly greeting. Cats often greet each other by sniffing noses, and the dogs in the study were observed doing this with cats. These nose-to-nose greetings occurred more frequently in the animals who had been introduced at a young age, suggesting that early exposure enables them to learn the other species' communication signals.

Another study, published in *Journal of Veterinary Behavior*, found that in general, dogs and cats living in the same house are friendly towards each other—but the experience of the cat is most important in mediating this relationship.[11] Early introduction of the cat to the dog (preferably before the cat is 1 year old) helped them to have a good relationship, whereas the age of the dog at first introduction was not important. (This is different from the previous study.) The most important factor in a good canine-feline relationship was the cat being comfortable in the dog's presence. It also helped if the dog was happy to share their bed with the cat, although cats were generally not willing to share their bed (perhaps because cat beds are smaller). Cats generally would not share their food with the dog or take toys to show to the dog, but it was a sign of a good relationship if they did. Cats and dogs also got along better if the cat lived indoors, perhaps because they got to spend more time together and learn about each other. But remember they should not be forced to interact; instead, let them choose whether or not to hang out together.

The relationship between dogs and cats was generally not described as close. For example, it was quite rare for them to groom each other. The scientists say that because cats have not been domesticated for as long as dogs, cats may find it harder to be comfortable around other animals. But the scientists also point out that dogs can be a real risk for cats, as dogs may try to eat them, whereas cats are unlikely to cause serious harm to a dog. Although aggression was most often reported from the cat towards the dog, it was likely because the cat felt threatened. A fifth (20.5%) of cats and 7.3 percent of dogs were said to be uncomfortable in the other's presence at least once a week. Two-thirds (64.9%) of cats and most (85.8%) dogs were said to be rarely or never uncomfortable in the other's presence. The study relied on owners' reports of their dog's and cat's behavior and did not make independent observations—something for future research, especially as people are not always very good at recognizing signs of stress in their pets. The results suggest that if your cat and dog are not friends, you should put extra effort into helping your cat feel comfortable around the dog.

The synthetic pheromones Feliway Friends (Feliway Multicat) and Adaptil may help cats and dogs to get along, according to a study that randomly allocated one of these to people who thought their cat and dog could use some help.[12] Both led to improvements, according to reports from the owners, and reduced problematic behaviors such as the cat running from the dog and the dog giving chase, the dog barking at the cat, and the cat hiding from the dog. Spending more time in a room together and greeting each other in a friendly manner happened more often with the Adaptil plug-in. So while the one to pick would depend on your circumstances, if you need your dog to calm down around the cat, Adaptil may be the better choice. A placebo effect seems unlikely given that Adaptil was only reported to make dogs more relaxed, and Feliway Friends to have this effect only on cats, but more research is needed on the use of pheromones.

When it comes to cats and other pets in the home, think about safety and remember that cats are predators. So if you have small animals like hamsters or gerbils, or any birds, you will need to keep them safe from the cat. As well, these animals may find it stressful if the cat is able to watch them in their cage. Some cats will also fish, so open-topped fish tanks are a risk and a fully enclosed aquarium will keep the fish safer. After all, fish is food to a cat, and food is always important, which we'll get to in the next chapter.

APPLY THE SCIENCE AT HOME

- Take a walk around your home and look at everything from your cat's perspective. If you have multiple cats, make sure they don't have to compete with each other for any resource, and that resources aren't located where one cat can block access.

- If introducing cats to each other, take it very slowly and only move forwards when both cats are happy. Don't force cats to interact with each other or with the other's scent; always give them a choice.

- Pay attention to the social relationship(s) between your cats. Remember to look for subtle signs of issues such as blocking as well as for overt signs such as play.

- If you want to have a cat and a dog, get the cat first if at all possible. The most important thing is to ensure that the cat is safe and feels safe. Make sure they have lots of high-up spaces and hiding places where they can get away from the dog, and use pet gates to keep the dog separate at times as needed.

10

FEEDING YOUR CAT

· · · · · · · · · · · · · · · · ·

W HEN HARLEY FIRST came to live with us, he was very particular about his food, especially how much he was fed. If
there was even a tiny bit of the bottom of his bowl showing, he would become very distressed, howling and howling and
howling. He clearly felt insecure about food, and so I would move
the bits of kibble so the bottom of the bowl was no longer visible,
topping it up if necessary to keep him happy. We had been told
that he was left behind when someone moved house, and taken to
the animal shelter some days later by a kind neighbor who heard
his howls. So he may have gone without food for a little while, or
perhaps he just had insecurities for another reason. It took about a
year before he stopped howling at the slightest glimpse of the bottom of his bowl and seemed to finally trust that I would feed him.

These days, Harley and Melina know their routine for being fed:
different food toys at different times of day. I put the food toys in

separate locations so they can eat without having to compete with each other. Melina always appears when it's time for food but does not ask for it at other times, whereas Harley . . . I expect you've already guessed that Harley will start to meow as his mealtime gets close.

At first, Harley did not like food that was different from his usual fare. If we offered little pieces of cooked chicken or tuna, Melina would gobble them up whereas Harley would sniff them and look at them as if to say, "What's this?" He didn't seem to know what to do with these "treats." He would watch Melina eating hers, and then she would rush in to grab his before we knew what was happening (she's fast, is Melina). I started pulling bits of chicken into tiny, tiny pieces and standing guard over Harley while he investigated whether he wanted to eat them or not. Now, some years later, he will eat some additional foods so long as the pieces are tiny, but he still prefers regular kibble or chicken pâté cat food, thank you very much. Melina, however, will eat almost anything, although the other day she did turn her nose up at a particularly stinky (and very tiny) piece of Camembert. If she smells chicken or tuna, she comes running from the other end of the house, meowing plaintively.

The way we are supposed to feed cats has changed to reflect our increased knowledge about normal feline behavior and what is best for cats. As well, there has been an increase in overweight and obesity in pet cats, which has implications for feline health. And getting things right with food can help to reduce stress for cats.

FEEDING CATS

THE NEED FOR cats to have multiple and separate key resources was discussed in chapter 3. Keep food and water away from the litter boxes. Instead, place them in quiet locations where the cats will feel safe eating and drinking (so the kitchen, although convenient for you, may not be the best place from the cat's point of view, especially if it is often busy). Wash food and water bowls every day (something many cat guardians don't do) to keep them clean and to avoid buildup of bacteria and biofilm (a.k.a. slime), and also because cats are picky eaters.[1] If you have multiple cats, they would prefer to eat separately. Remember, they are solitary hunters and would normally hunt and eat on their own.

If need be, you can put one cat's meals in locations other pets cannot reach or squeeze into, and/or use automatic feeders that will open only for the correct animal's microchip. Some cats prefer still water, whereas others prefer moving water, such as from a dripping tap or a special water dispenser made for cats. Keep an eye on how much your cat is drinking and eating; if you see any changes, always consult your vet.

The American Association of Feline Practitioners recommends that a cat's daily calorie count (roughly 40–66 kcal per kg per day) be split into multiple small meals throughout each twenty-four-hour period.[2] International Cat Care suggests you feed your cat five (or more) small meals a day. The reason for many small meals is that it suits cats better. A feral cat who had to catch mice for breakfast, lunch, and dinner might catch and eat ten to twelve mice a day. A mouse is only a small meal, so this amounts to multiple small breakfasts, lunches, and dinners. Ten meals a day is not very realistic for cat guardians, but five is much more achievable.

Since it is likely you will be out for some of those meals, you can use an automated feeder and set the timer to provide meals. For most cats, it's a good idea to put the food bowl or food toy somewhere high up, because cats like high-up places. However, for cats with arthritis or other mobility issues, you may need to provide the food closer to the ground or provide steps to get there so they don't have to jump.

Puzzle feeders are toys that make the cat work to get the food out. They provide enrichment for your cat by making them use their skills to access the food. Maybe the cat has to stick their paw in and pick pieces of food out, or maybe they need to roll the toy around with their nose or paw to make food fall out of the holes. Some food toys stay in one place, and others, the cat has to move around. Most are designed for dry food like kibble, but some also work with wet food. There are plenty of commercially available food-puzzle toys to choose from, but it's also very easy to make your own. For wet food, for example, try putting small portions in cupcake holders or bowls and hide them around the house. Your cat will be able to use their nose to sniff them out. You can find ideas and reviews at a website called Food Puzzles for Cats run by Dr. Mikel Delgado and Ingrid Johnson.

When your cat is new to food puzzles, you need to make the toys easy to find and operate and use treats to get your cat interested. Cats may go on strike if the puzzles are too hard to use, and not eating is very dangerous for a cat. So start with easy ones. Fill them to the brim to make it easy for your cat to tip treats out, and make sure any openings are on their widest settings, that there are multiple holes for food to fall out of, and that the food inside is very tempting. As well, if the toy needs to be moved over a surface, make sure it is on an easy surface—for example, a linoleum or hardwood floor rather than carpet or a deep-pile rug. Over time, your cat can gradually progress to more difficult food toys, but you want to be sure their first experiences are good ones. Again, multiple cats in the same home should not have to share food-puzzle toys, although cats that get on well may be happy to do so.

Only around 5 percent of cats have food-puzzle toys, according to a study published in *Journal of Applied Animal Welfare Science*, although I think they are becoming increasingly popular as more and more people hear about them.[3] Food puzzles make cats engage in part of their natural predation sequence—getting food. Using these skills has many benefits, according to a report in *Journal of Feline Medicine and Surgery*, including encouraging cats to be more active, reducing their stress levels, and making them less demanding of their owners.[4] Dr. Mikel Delgado, one of the authors of the report, told me, "It's a great way to give your cat something to do to keep them busy and get them doing what a predator is supposed to do ... Working for their food!! It's great for their brains and body. A bonus is that it's really fun to watch your cat play with a food puzzle!" Switching to providing food via puzzle toys is often a part of the solution to behavior issues, as it gives the cat something to do.

One time when we still had Bodger, I was working at home and heard a strange noise from the kitchen. I went to investigate and found that Melina had knocked a packet of dog treats (freeze-dried liver) off the kitchen counter where we kept it for easy access, had scratched holes in the packet with her claws, and was flinging it around to make the treats fall out. She had basically made her own food toy! ("You taught her to do this," said my husband.) We had to start keeping the dog-treat packet in a cupboard and fastening the cupboard door.

The scientific term for working for food by choice is contra-freeloading. This approach to feeding is commonly practiced with zoo animals as a way of providing enrichment in their enclosure and is also becoming increasingly popular for pet dogs. It is recommended for pet cats as a form of enrichment and also to help resolve behavior issues. Again, take care to introduce food puzzles

at a very easy level and use tasty treats to help get your cat interested in them. Other ways to encourage foraging include dividing
the cat's food into portions and hiding them around the house,
including in high-up places.

Cats are what is known as obligate carnivores: they have to eat
meat. They need their food to contain arginine and taurine (amino
acids without which they become ill), vitamin A, and vitamin D
(because they cannot make enough).[5] If you would prefer to feed
your pet a vegetarian or vegan diet, then a cat is not the right pet
for you. Commercially available cat foods are designed to meet
cats' nutritional needs and their differing tastes.[6] Some cats prefer the foods they ate early in life, although they can learn to eat
other foods, and sometimes cats can get bored of a particular food.
Cats who are finicky eaters are sometimes stressed, or they have
trained their people to offer better foods by rewarding them with
affection. Although one study found that cats fed food with a high
meat content (no grains, no meat meal, and no rendered meat)
hunt less, it is not known if a specific ingredient or other aspects
of the food (such as its novelty) better satisfied the cats.[7]

Kittens should have a food that is designed for kittens, because
they need calorie-dense food. In the United States and Canada,
look for a label from the Association of American Feed Control
Officials (AAFCO) saying that it is suitable for growth or for all life
stages. From the age of 1 year, cats can have adult food. Although
some pet foods are marketed for senior cats, one study found
they contain more fiber but otherwise have few differences from
other types of commercial cat food.[8] Therefore, the scientists recommend that you buy food based on your individual adult cat's
preference rather than their life stage. Talk to your vet if you need
guidance. If your cat has a specific condition, such as chronic kidney disease or dental issues, your vet may prescribe a therapeutic

diet. Treats and supplements do not need to be nutritionally complete, but if your cat needs a special diet you will need to ensure the treats also meet those dietary needs.

Some people foods are toxic to cats. The ASPCA says you should not feed your cat chocolate or anything containing caffeine; alcohol; avocado; macadamia nuts; yeast dough (because it will rise in your pet's stomach and can even cause a rupture); raw or undercooked meat, eggs, and bones (due to potentially toxic bacteria); anything containing the artificial sweetener xylitol (e.g., some peanut butters); onions, garlic, and chives (which cats are even more susceptible to than dogs); or large amounts of salt.[9] Milk products can cause diarrhea, but specially formulated cat milk is available if you want to give it to your cat.

"One thing that would make a better world for cats is owners understanding better the role food plays in their relationship with their cat. For many owners, food is a currency of love. Offering excess food and especially highly palatable food not intended for cats can harm their health. Helping owners to understand their cat's need for the appropriate amount of a complete and balanced diet is key to tackling the problem of feline obesity, now an issue for almost 60 percent of US cats. We need to help cat owners change their behavior without diminishing their bond."

—SANDRA MCCUNE, PhD, visiting professor, Human-Animal Interaction, University of Lincoln

OVERWEIGHT AND OBESITY IN CATS

MANY CATS ARE overweight or obese. According to a review published in *Journal of Feline Medicine and Surgery*, an obese cat is almost four times as likely as a normal-weight cat to develop diabetes, and more likely to suffer from other medical problems such as urinary tract disease and lameness.[10] Just ten extra pieces of kibble a day, over and above what the cat needs, will cause a 12 percent increase in weight over the course of a year. A typical cat should weigh about 4.5 kg (10 pounds), but many people are not good at recognizing when their cat is overweight. In the review paper, comparisons are made to human weights to drive home how serious an issue it is. For example, a cat who weighs 6.8 kg (15 pounds)—about 50 percent more than they should—is equivalent to a weight of 98.9 kg (218 pounds) for a 1.63-meter (5'4") woman, or 115.2 kg (254 pounds) for a 1.75-meter (5'9") man.

Weigh your cat regularly to keep an eye on any changes. It's probably easiest to stand on the scale with and without your cat in your arms and then subtract your weight to find that of your cat. Apart from using an accurate scale, look at the shape of your cat. The waistline should be visible, as should a tummy-tuck, and you should be able to feel the cat's ribs. The World Small Animal Veterinary Association has a chart to show Body Condition Scores in cats on a scale from 1 to 9, where 5 is a perfect weight.[11] As a cat becomes overweight and then obese, the waist becomes less obvious or not visible, and the ribs become harder to feel and then cannot be felt. A common finding from studies of overweight and obese cats is that their owners have underestimated how overweight they are. If you're not sure whether your cat is overweight, ask your veterinarian.

The Bristol Cats Study, which follows pet cats throughout their life from kittenhood, found that by just over 1 year of age,

7 percent of cats were already overweight or obese, according to their owner.[12] Two main risk factors were identified. Cats who were indoors only were more likely to be overweight, as were cats fed more than half or all of their food as dry kibble. It is especially unfortunate for a cat to be overweight at this young age, because overweight cats are likely to become obese. An American study also looked at the risk factors for overweight and obesity in pet cats.[13] Amongst a very large sample of cats who had seen veterinarians around the United States, 35 percent were found to be overweight or obese (with a Body Condition Score of 3.5–5 on a 1–5 scale). This number rose to 40 percent for cats aged between 5 and 11.

The risk factors for being overweight were being male, being neutered, and eating a therapeutic or premium food. Overweight cats were at increased risk of oral disease and urinary tract disease. The breeds that were more likely to be overweight or obese were Domestic Shorthair, Domestic Mediumhair, Domestic Longhair, and Manx, as well as cats of unknown breed. As well as being at increased risk of oral disease, cats who were obese were also more likely to have cancer, diabetes mellitus, and skin diseases. In this study, relatively few old cats were overweight or obese, but it is not known if this is due to cats losing weight as they get older or to overweight and obese cats not reaching those ages. A study of pet cats in Paris also found similar risk factors of being spayed or neutered and being fed a therapeutic diet.[14] In that study, sometimes the cats were fed little bits of meat or milk, but it wasn't possible to tell if this was a risk factor or not. Therapeutic and premium foods are often much more calorie-dense than kibbles sold in supermarkets.

Special diets are often given to cats who are overweight or obese. These diets contain more nutrients per calorie so the cat does not miss out on nutrients when the amount of calories they are fed is reduced. One study looked at overweight and obese cats in twenty-five different countries who were put on a three-month weight-loss diet.[15] The number of calories the cat should have each day was calculated, and owners in most countries used a scale to measure the food in grams, while those in America were given a special cup measurement. They were also told to split the food into at least two meals a day. Although 710 people agreed to start the trial, only 426 cats took part for the full three months. This perhaps shows how difficult it is for people to help their cat lose weight, although the reasons cats stopped taking part were not always known.

At the start of the three months, the cats on average had a Body Condition Score of 8 (on a scale up to 10) and an average (median) weight of 6.7 kg (14.8 pounds). The good news is that 97 percent of the cats lost weight by the end of the three months, and 5 percent of them had even reached their target weight. According to their owners, 49 percent of the cats were more active, which is also good news. The survey asked owners about the cat's quality of life, which was said to be the same in 49 percent of cats, improve for 38 percent, and get worse for 12 percent. For about half of the cats (48%) food-seeking behavior decreased, but it got worse for 18 percent of the cats and was unchanged for the rest. One other thing to note from this study is that weight loss was greatest at the beginning but then slowed. This may be disheartening for cat owners to see, but it was still worth continuing with the program given the changes in the cats' weights by the end of the three months.

As well, even smaller amounts of weight loss can make a big difference to health. One surprising finding of this study that needs to be followed up is that the cats fed dry food only lost more weight than those fed wet food only or a mix of wet and dry food. This is especially surprising given that both were formulated as weight-loss diets, and other research has suggested that dry food is associated with weight gain. So we need more research before we can draw definitive conclusions.

Sometimes people are reluctant to put their cat on a diet, because they worry that their cat will not be as affectionate towards them. There's some good news on this front from a study of fifty-eight obese pet cats published in *Journal of Veterinary Behavior*.[16] Over a two-week period, the cats were transitioned to a new diet, and then they stayed on that new diet for eight weeks, with regular weigh-ins at a university vet hospital. Three different diets were tested: a control diet (designed for cats who do not need to lose weight); a high-fiber diet; and a diet low in carbohydrates and high in protein. The scientists calculated how much kibble each cat should have each day, and the cats' guardians were given a measuring scoop for the food.

First of all, almost every cat lost weight over the eight-week period on the calorie-restricted diet, regardless of the particular food they were fed. The cats on the high-fiber diet lost the most weight overall. Second, despite being on a calorie-restricted diet, the cats actually became more affectionate, according to their owners. There was an increase in begging, but people seemed to see this behavior as affectionate. One option the scientists suggested to reduce begging would be to use a feeding device that opens at a set time to let the cat have the food. With food coming from the device, the cats would be less likely to beg their owner for it. Interestingly, in this study, there were no differences in behavior changes between cats who had previously been fed ad libitum (food left out all day) and those who had been fed at set times.

Feeding behavior does change when cats are on a calorie-restricted diet, according to a study of group-housed cats that lived at a facility.[17] All of the cats were used to being fed ad libitum, then a group of them were switched to a restricted diet while the other cats remained on the same feeding schedule. The cats were fed like this for nine months, and those on the restricted diet lost weight during this time.

When the cats on the restricted diet were switched back to ad libitum feeding, they regained the weight. But there were also changes in how they fed: they ate more food at once, ate faster, and ate fewer meals overall. This is in contrast to the cats who remained on ad libitum feedings throughout the study, who ate more frequently, but smaller, slower meals. As well, the cats in the calorie-restricted group showed some signs of tension and aggression between them, which suggests the calorie restrictions may have made them irritable. These signs went away when they returned to the ad libitum feeding plan. Because of this finding, the researchers suggest that in homes with more than one cat, it is a good idea to feed them in separate areas when they are on a calorie-reduced diet. The scientists also recommend using puzzle feeders to increase activity around food and make each meal take longer, and to divide the day's food into separate portions to be fed throughout the day.

Some evidence suggests that owners of overweight cats have a tendency to overhumanize them, according to a study of 120 cat owners in Germany reported in *Journal of Nutrition*.[18] The study looked at cats who were indoors only or had access to an enclosed balcony/garden (no free range). Normal weight was defined as a queen under 4 kg (8.8 pounds) or a tom under 5 kg (11 pounds), and overweight as over 5 kg (11 pounds) or 6 kg (13.2 pounds) respectively. Owners of overweight cats had a closer relationship with their cat and were more likely to say the cat consoled and encouraged them. They were also more likely to say their cat was like a child to them. Although both sets of owners talked to their cats, owners of overweight cats were more likely to talk to their cat and to talk about topics relating to friends and family or work. Owners of overweight cats were more likely to watch their cats eat, suggesting that food played a greater role in their relationship. Owners of normal-weight cats were more likely to play with their cat.

Many people use approximate measures for food; an accurate cup measure is better, and weighing the food with an accurate scale is best of all. When scientists asked dog owners to measure out portions of kibble using either a one-cup food measure, a two-cup food scoop, or a two-cup liquid jug, they found the resulting amounts of food varied dramatically from too little to too much.[19] As well, although one in five of the participants had a weigh scale at home that they used for their own food, only one in fifty used it to measure out dog food. However, after having seen the errors in their measurements, most said they would be likely to use a scale in future. Since it seems likely that cat owners also make similar errors of judgment, and cats eat smaller amounts of food than most dogs, weighing the food is better.

Before starting your cat on any weight-loss program, be sure to speak with your veterinarian so you can develop a safe and effective plan. To lose weight, a cat needs around 60–70 percent of the calories required to maintain their weight. Cats must only lose weight very slowly. If a cat loses weight too fast, they develop fatty liver disease, as their liver cannot break down the fats fast enough, and if not treated in time this can be fatal. A cat that is a little overweight should have a fixed amount of regular food, measured out each day.[20] For very overweight and obese cats, your vet might recommend a special weight-loss diet to ensure they still get enough nutrients. If your cat is obese, a substantial weight loss that does not reach normal weight will still have health benefits. Try keeping a food diary of everything your cat eats to help you stick to the plan. Remember to include products designed for dental health, as they also have calories. And you can continue to give your cat treats, since it makes them happy, but reduce the amount of kibble to take account of the calories from treats.

Finally, remember that a plateau after initial weight loss is normal and does not mean that you and your cat are failing. You are helping your cat achieve better health, so celebrate the small successes along the way.

APPLY THE SCIENCE AT HOME

- Keep an eye on your cat's weight and speak to your veterinarian if you have any concerns about it.

- Try to feed your cat multiple small meals a day. If being out at work for part of the day makes this difficult, you could consider a timed feeder that will open at a set time.

- Feed at least some meals via food toys. Remember to start with very easy toys with treats to get your cat's attention. You may have to help them at first, but they'll get the hang of it. You can gradually progress to harder toys.

- Measure your cat's food with a weigh scale, rather than eyeballing it or using a cup measure. This will help you be more accurate in the amounts you feed your cat.

11

BEHAVIOR PROBLEMS IN CATS

· · · · · · · · · · · · · · · ·

THE SMELL WAS unmistakable. I followed my nose down the hallway, into the dining room, and over to the corner. The brown carpet doesn't show much, so I dabbed with a paper hankie to check the area. The hankie came away damp and a very dark yellow heading towards brownish. I set to blotting up the pee, putting an enzyme cleaner on the affected area, and then after a short pause dabbing that up too. Now the house smelled okay again, but I had a dilemma: Which of our two cats needed to go to the vet? Later that day I saw Melina head over to that same spot and squat to pee. I knew better than to yell at her. I pretended not to notice and then as soon as she'd gone away, I cleaned up the new mess. Now I knew which cat I had to get a pee sample from.

Because of the layout of our house, and the fact that at that time we had to keep one of our dogs (Ghost) separate from the cats, it wasn't easy to shut Melina in a room on her own. Instead I set up the biggest dog crate—which was very large—to make it comfy for her. And in the litter tray, instead of regular litter, I put the little black plastic pellets the vet had given me. They made a sorry little pile at the bottom of the tray, leaving lots of empty space where the litter would normally be. The aim was to use the unabsorbent litter so that I could collect a pee sample. So I put Melina in the crate and waited. And waited. And waited. Poor Melina wasn't very happy. Late in the evening I saw her look at the litter box a few times and go away from it. But eventually she peed. Then I let her out of the cage, collected the urine sample in a tub, and put it in the fridge.

The next morning at the vet, a urine test showed that she had an infection. Melina was prescribed antibiotics and because her urine was so concentrated, I was advised to add some water to her food each day. Luckily, she was still happy to eat her wet food with a bit of water in it, the antibiotics resolved the infection, and she went back to toileting in the litter box. I also put some extra water bowls around the house in case they might be used, one of which became part of the regular setup of the house, because she does occasionally drink from it.

The experience of having a cat toilet outside the litter box is not unusual. Unfortunately, many people assume their cat's mishap is due to spite or hatred, don't think to take their cat to the vet, and, in some cases, end up rehoming or euthanizing their cat. House soiling is one of the most common behavior problems reported by cat guardians. Amongst cats recently adopted from shelter or rescue, the most common behavior issues reported soon after adoption are inappropriate scratching or chewing of furniture and house-soiling issues, according to an Australian study reported in *Applied Animal Behaviour Science*.[1] In another study, cat guardians reported fear and anxiety as the most common behavior issue, followed by destructive behavior (e.g., scratching), house soiling, excessive vocalizations (including at night), and aggression.

Some behavior problems arise because the cat's guardian has not provided things the cat needs in their environment—such as adequate scratching posts or litter boxes.[2] Other problems arise due to stress, which may be caused by an inadequate environment but also by the guardian's use of punishment, the presence of other cats who aren't in their social group (or cats seen through the windows), inadequate socialization, or other factors. And other behavior problems arise because of medical issues, including pain. Pain may be a factor in a wide range of behavior issues, including aggression, attention seeking, pica (eating strange objects), and house soiling (for example, the cat may find getting in and out of the litter box painful or may try to avoid going up or down stairs to get to it).[3] It is also possible that pain may influence other issues by making the cat more fearful, anxious, or pessimistic. The cat guardian's knowledge (or lack of knowledge) of cat behavior can be a reason behavior issues develop and also affect how those issues are resolved.

One study reported in *Animals* found that cat guardians with a better knowledge of cat behavior were less likely to have a cat with behavior issues, and less likely to use positive punishment to attempt to resolve any issues that did exist.[4] Unfortunately, using positive punishment can cause stress and affect the cat's bond with their person (see chapter 5 for more on training). To help you understand more about your own cat's behavior, we'll consider some of the more common behavior issues and what can be done about them. Whenever a cat's behavior changes, it is important to get them checked out at the vet in case of a medical cause or contributing factor.

HOUSE-SOILING ISSUES

HOUSE-SOILING ISSUES CAN be very upsetting for the people in the home, but undoubtedly also for the cat who may have a painful medical issue and/or be very stressed. Never punish your cat for house soiling as it will only stress your cat, potentially making the problem worse and also affecting the human-animal bond. Urinating outside the litter box is known as periuria and is considered to include both marking (also known as spraying) and inappropriate urination (latrining). Marking is using urine as a chemical signal for other cats, and often involves a vertical stance (although not always) and a vibrating of the tail, whereas latrining is typically a squat stance to pee. Always take your cat to the vet for house-soiling issues, so they can be checked for a medical problem. Your vet will advise you on the best way to collect a urine or stool sample as needed.

"Before it becomes a behavioral problem we need to rule out medical problems," says veterinary behaviorist Dr. Wailani Sung. "It's not spiteful even though the cat may seem like it is being so. Typically when people think the cat is spiteful it's because the cat not only eliminates outside the litter box but they do it in areas that the owner is offended by. They do it on their favorite couch or they do it on their pillow or on their bed. And when they do it on their pillow and bed, the owners take it very personally, like 'They're being spiteful,' right. Or, 'I said no and I locked the cat out and I come out of the room and the cat had peed on my favorite blanket or my pillow.' So that's why they think it's spiteful, whereas I see it as a cry for help. I see it in that particular situation, the specific location where the cat has urinated or defecated is a cry for help, and it's an indication that something is wrong."

She explained that if cats are in pain when they toilet, they often associate the pain with the litter box. "There are many underlying causes for cats not using the litter box, and if cats are uncomfortable in their bladder or their colon they kind of have this weird association with the litter box. 'Every time I go and eliminate, it's happening in the litter box, so maybe it's related to the litter box or the specific location.' Like shoes and other locations, and they keep trying to find different places. It's really important that the owners rule out underlying medical problems like bacterial cystitis, which is a bladder infection. They also need to rule out cystic calculi, which is bladder stones. And then some cats, when they're anxious, are prone to getting feline interstitial cystitis. It's basically a sterile cystitis and they might have blood in their urine or their bladder but there's no bacterial cause for that. So that's usually related to anxiety and it's related to the whole stress complex related to the sympathetic nervous system and the HPA axis. It manifests in a lovely way in the bladder of the cat." The hypothalamus, the pituitary gland, and the adrenal glands work in concert (the HPA axis) when an animal (including humans) is stressed.

If the vet rules out a medical issue, have a look at the toileting space. I asked Dr. Sung to summarize what to consider. "It's usually determining if you're giving the appropriate amount of litter, type of litter box, the size of the litter box, the location, and cleanliness of the box," she said. "Those are probably the top five I immediately start off with. And then we ask about other factors. Where exactly is it located? Is it in a high-traffic area? Are there other animals in the house that are disturbing the cat, or other family members that might be disturbing the cat?" She explained that cats like their privacy when they are toileting. Cats are prey animals for larger predators and so they are always on the lookout to make sure they are safe. "They have some of that prey-like behavior where when they need to eliminate, the few seconds of them not being able to move—I want to say off the top of my head it's probably between like anywhere from ten or twenty seconds of having to do their business—and it's really hard to not get upset when someone's moving and you just want to sprint away. So if they get disturbed all the time, they are not going to want to use that area anymore."

The importance of environmental factors is shown by the results of a study that looked at 294 cats who were surrendered to a shelter for peeing and/or pooping outside of the box.[5] In the shelter, the cats were given a choice of litter boxes and litter types until they found one they consistently used, and they were housed on their own, away from other cats. These cats were adopted out at a similar rate to cats who had been surrendered for other reasons, and although they had a slightly higher rate of being returned to the shelter, this was mostly not due to house soiling. So this study shows that, after seeing your vet about potential medical causes, making environmental changes and reducing stress can be effective. It also shows that cats with house-soiling issues can be successfully adopted and should not be euthanized for this reason.

To ensure your cat is happy with their litter box setup, make sure you provide enough litter boxes and that they are large enough. Use a litter substrate that's a friable material and that's easy for the cat to scratch at; it should not be strongly scented. Don't use litter liners, because a cat's claws can easily get caught in them and because it may make the base of the litter tray slippery. Scoop the litter tray once or twice a day, and dump and clean out the whole thing and fill it with new litter once a week, or sooner if it becomes soiled.

Although some people like to train their kitty to use the human toilet, it's not a good idea. For one thing, the cat risks falling in, and they may find it trickier to jump up and balance on the seat as they get older and potentially have arthritis or other issues. The other reason not to encourage this practice is that you want to keep an eye on what comes out of your kitty, as changes in volume or signs of blood may indicate a health issue for which they should see the vet.

Every cat is an individual and will have their preferences. Some cats prefer to urinate in one tray and defecate in a different one. When presented with three identical litter boxes, one containing a wood-pellet litter, one with clay granules, and one with silica microgranules, the least preferred (i.e., least used) option was the wood pellets, according to a study of twelve cats published in *Journal of Veterinary Behavior*.[6] In a follow-up study, another twelve cats were given a choice between two identical litter boxes that contained granule-size litter, one made of silica and the other of clay. In this case, cats preferred the clay litter (i.e., used it more than the silica litter). This suggests that cats prefer clay granules as a litter substrate, although it was a small study and there are many other types of litter available that scientists have not put through the same kind of test.

Cats prefer a larger litter box, according to a study in *Journal of Veterinary Behavior* that gave people from forty-three households two litter boxes to use during the study period. One of the boxes might be called "regular" in size, and one was 86 cm (33.9 inches) in length, longer than most commercially available ones.[7] The cats' owners were given clumping cat litter and instructed to keep a record of all of the deposits, whether pee or poop, over a four-week period. For the first two weeks, the boxes were set in particular locations in a room, and then after two weeks, the litter was completely replaced and the locations were swapped over for the remaining two weeks. Over the four-week period, the seventy-four pet cats who were taking part made more than 5,000 deposits in the large boxes, and just over 3,200 deposits in the small boxes. Interestingly, for the first two days after the boxes were switched, the cats had a location preference, but this effect then faded away. This study shows that cats prefer a larger box when it is available to them, although they did also use the smaller box.

As to whether cats prefer a covered or uncovered litter box, it seems to depend on the individual cat. One study published in *Journal of Feline Medicine and Surgery* gave twenty-eight cats a choice between a covered box and an uncovered box that was otherwise the same, including being scooped at the same times.[8] The results showed that although some cats have preferences, many do not and are equally happy with both types. Another study, this time in *Journal of Veterinary Behavior*, tested twenty-five cats and found a preference for a covered box—but only when this box was also larger.[9]

What really matters to all cats is cleanliness. In one enterprising study of multi-cat households, published in *Behavioural Processes*, researchers added either the smell of pee and/or poop or odorless fake pee and/or poop to cats' litter boxes.[10] The added smell did not seem to trouble the cats, and they did not seem to mind whether it was their own or from another cat in the household. However, they did not like the added lumps of fake pee or poop, especially the poop. This finding shows the importance of regularly scooping the litter box.

Cats with house-soiling issues spend less time digging in the box compared with cats who have no such issues, according to a study published in the *American Journal of Veterinary Research*, but there were no differences in the amount of time spent sniffing, pawing, or covering what they had just done.[11] This study also did not find any differences in the location of litter boxes in the home between the cats with and those without elimination issues.

Another study gives us more clues as to when a cat may not be so happy with their litter box. The study, published in *Applied Animal Behaviour Science*, looked at the behavior of twelve cats in two different environments: one set up as an enriched environment with a large litter box filled with sandy litter, and another that was an impoverished environment more like a clinical setting with a small litter box filled with little polypropylene beads.[12] The researchers looked at behavior before, during, and after elimination (both poop and pee) and found thirty-nine different behaviors that are potentially associated with elimination in cats. It seems that cats will still use a litter box they are not entirely happy with, but it will take them longer to eliminate; they may be reluctant to go in it, or may go in and out of it a few times; and may even keep one paw out of the box altogether. In contrast, in the enriched environment with the large litter box, cats went in, did their business, and that was pretty much it. Another finding was that cats would pee less often and with bigger pees in the small, clinical litter box. This is important because holding urine even when wanting to go might lead to the development of medical issues. In fact, it took the cats fifty-two seconds, on average, to pee in the small litter box (i.e., the duration of urine flow), compared with twenty seconds in the large litter box.

Another big difference occurred in what the cats did after peeing or pooping. In this clinic-like setting, they spent much longer pawing at the area around the litter box—the floor, the wall, the sides of the box. As well, they spent much more time sniffing their deposit (just under a minute sniffing poop, and just under thirty seconds sniffing pee). In some cases, they went back to sniff it again several times. In this litter box, they were not able to cover up their pee or poop, because there weren't enough little plastic beads to do so. What this tells us is that scent plays an important role in the cat's preferences for toileting. If you notice that your cat seems unsure of going in their litter box, goes in and out again, takes their time over peeing, or spends a long time pawing or scratching afterwards, it is worth considering whether you can improve their toilet setup.

The litter box was not associated with some types of house soiling in a survey published in *Frontiers in Veterinary Science*.[13] In this study, urinating outside the box was more often diagnosed as marking if it was on a vertical surface, the cat stood up to eliminate, there was only a small volume of pee, and the cat did not try to cover the area afterwards. Marking was less likely to happen with cats who were described as having a relaxed personality. Marking was six times more likely in a multi-cat household compared with a home with only one cat, and was also more common with older cats, if there was a cat flap, and if the cat had outside access. Latrining was more likely when the cat did not have outside access, when the cat was also pooping outside the box, and with cats who were very clingy. Latrining was also more common in a multi-cat household.

Subsequent research, published in *Journal of Feline Medicine and Surgery*, found that common medical issues in cats who either spray (mark) or urinate outside the box include poor kidney function, cystitis, bladder stones, and glomerulations (little hemorrhages in the bladder wall).[14] As well, this study found that if one cat in the household had a medical issue related to their latrining, other cats in the household were equally likely to have a similar medical problem even if they were toileting normally. In contrast, the housemates of cats who were spraying were not likely to have a medical issue.

The American Association of Feline Practitioners and the International Society of Feline Medicine provide guidelines for working through house-soiling issues with cats.[15] As mentioned, always begin by taking your cat to the vet. Often, even when the vet diagnoses a medical issue, you will still need to make changes to the litter boxes and/or how often you scoop them, or more generally to reduce stress for your cat. Sometimes it can be challenging to resolve these issues and so it helps to have your vet or a cat behavior counselor provide some guidance and encouragement along the way. Above all, remember the cat is not acting out of spite; instead, it is a medical issue, a stress issue, or a complicated mix of both.

SCRATCHING ISSUES

SCRATCHING IS A normal behavior for cats, so they must have places to do it. If they are scratching in places you'd rather they did not, then you must provide them with suitable surfaces that they can scratch, bearing in mind that cats may have individual preferences. It is best to have both horizontal and vertical scratching posts, and if you have more than one cat, you need to have enough scratching posts for each cat. See chapter 3 to ensure you have the right kind of scratching posts, and reward your cat with a treat or some petting (whatever your cat prefers) for using them. Remember that punishment is not a good idea, because it can cause fear, anxiety, stress, and a worse bond with you.

Don't declaw your cat. Where I live, this surgery is banned, but unfortunately in the United States it is still allowed, with the exception of New York State. At time of writing, both Florida and Arizona lawmakers are considering legislation to ban declawing, and two major veterinary chains (Banfield and VCA) as well as Fear Free–certified practices will no longer declaw cats. The word "declaw," although commonly used, is a misnomer. The technical name is onychectomy, and it means amputating the knuckle along with severing the tendons, nerves, and ligaments. It is a painful procedure that takes the cat weeks to recover from and has long-term consequences. Without claws, the cat is no longer able to defend themselves; and after the amputation, cats may find using the litter box painful.

One study reported in *Journal of Feline Medicine and Surgery* found that declawed cats were much more likely to have back pain and to have the behavior problems of peeing and pooping outside the litter box, excessive grooming, and aggression.[16] When declawed cats toileted outside the box, it was typically close to the litter box but on a soft surface such as carpet or fabric, which suggests they found touching and digging the cat litter painful. A vet exam also found signs of pain and that bone fragments were commonly left behind during the declawing procedure.

Remember, declawing does not benefit the cat, and surely we should have our cats' best interests at heart. The Paw Project, which campaigns against the declawing of cats, points out that cats' paws cannot ever be returned to normal.[17] If your cat has been declawed, it is important to talk about the possibility of pain with your veterinarian.

"One thing that would make a better world for cats:
having a regulatory body for animal behaviorists and trainers
to ensure animal welfare standards are met using up-to-date
evidence-based methods. While this may not be the first thing
that springs to somebody's mind for improving cat welfare,
ultimately this has the potential to make the biggest impact on
cats around the world. In the UK, we have the Animal Behaviour
and Training Council. It sets and maintains the standards of
knowledge and practical skills needed to be an animal trainer,
training instructor or animal behaviorist, and it maintains the
national registers of appropriately qualified
animal trainers and animal behaviorists."

—NICKY TREVORROW, behavior manager at Cats Protection

FEAR AND AGGRESSION

FEAR AND ANXIETY are common and can cause aggression. If a cat
is afraid, they would prefer to escape and hide; however, if this is
not possible, they may become aggressive in an attempt to keep
people away. You might see this behavior at the vet clinic (one
reason to choose a vet who specializes in cats and/or Fear Free or
Low Stress Handling), or when visitors to the house try to pet a
fearful cat who has no escape route, or if the cat is afraid of children in the home. And sometimes when the cat is in a high state
of arousal, they will redirect the aggression to the nearest target,
which might be you.

Another cause of aggression in cats is frustration, often because
they are bored or cannot get something they want. When you are
petting a cat and they seem happy and then suddenly turn on you

or even bite or scratch, they are often frustrated. In this case, the petting likely went on for too long and became too much. Try to keep petting sessions frequent but short, and watch for signs of increasing arousal such as a swishing tail or rippling skin. Sometimes cats in shelters can be aggressive due to frustration and will tip their litter tray and food bowl over, and may attack staff trying to leave the room they are in. Here, the solution includes making sure the cat's environment is right for them and giving them plenty to do, including play with toys such as food-puzzle toys and wand toys.

Both genetics (nature) and life experiences (nurture) play a role in why fear develops. In one study published in *Physiology and Behavior*, thirteen litters of kittens were divided into three groups.[18] One group acted as a control group that was treated normally. The kittens in the other two groups were weaned early and either handled or not. Every four weeks between the ages of 8 and 20 weeks, the kittens were tested to see how friendly they were to people. The results showed that both genetic factors (how friendly the father cat was) and environmental factors (the handling regimen) influenced the kittens' behavior towards people.

Another study, published in *Applied Animal Behaviour Science*, found similar results.[19] This time, researchers divided the kittens into two groups. Half of the kittens in each group were known to have a friendly father, and half to have a father who was not friendly to people. When the kittens were aged 2–12 weeks, people handled the kittens in one group but not the other. Researchers then tested the kittens to see how friendly they were to a person they knew and to a stranger, and how they acted towards an object they had not seen before. The kittens who had a friendly father and were handled during and after the sensitive period were the friendliest. The kittens born to a fearful, unfriendly father were quite friendly if they had been handled, but not as friendly as the kittens in that group with a friendly father. The kittens with an unfriendly dad and who were not handled were slower to go to a new person and did not interact with them as much. They were also slower to approach the new object and did not interact as much with that either. The results of this study and the previous one suggest that you should find out how friendly a kitten's parents are before bringing the kitten home.

A kitten's response to stress is also affected by the epigenome, the chemicals that bind DNA. While it used to be thought that this stress response was reset when genes are passed on, it turns out that some epigenetic changes can be inherited. When rat moms do a good job of nursing, licking, and grooming their pups, their pups grow up to be calm adults who are resilient in dealing with stress, but if rat moms are not good at nurturing their pups, those pups grow up to be anxious adults.[20] This is due to epigenetic change, not genetics.

Adaptations to a stressful environment occur through epigenetic processes called histone modification (which affects how easy it is for parts of the gene to be transcribed) and methylation (which effectively turns genes off). If the kittens are born into a stressful environment, these epigenetic changes allow them to adapt more easily. When the kittens have these changes but don't live in a stressful environment, it can be maladaptive and cause issues such as anxiety.

Prenatal stress can also affect the development of hormones and the nervous system in the developing fetus, and these changes can continue after birth. As well, if the mother is stressed, she may not be able to take such good care of her young. When scientists stressed mother cats by not giving them enough to eat, the kittens had poor balance, interacted less often with their mother, and developed more slowly in the two to fourteen days after they were born.[21] Therefore, if you ever have a pregnant cat, make sure she has a stable, predictable environment that she has some control over and enough high-quality food to eat. She will benefit, and her kittens are much more likely to develop normally.

Experiences during the sensitive period for socialization and the weeks and months after, as discussed in chapter 2, are another important factor. Fear can be instilled simply from not having a wide enough range of positive experiences and handling during this time. That's why it is important to check that kittens come from a home where they have been handled and experienced normal household life (rather than being kept in a barn, for example). And this is also why you should continue to socialize and habituate your kitten when you bring them home. As well, bad experiences at any time of life can induce fear. Although a cat (or other animal, including humans) needs some fear to stay safe, too much fear, or ongoing stressors that lead to prolonged anxiety, not only can cause behavior issues but also are bad for your cat's health.

When scientists surveyed the owners of young cats (1–6 years) who had been adopted from a shelter where they were fostered as kittens, they found that aggression was not linked to early management (perhaps because the foster homes were all good ones).[22] Female cats were more likely to be reported as aggressive to both people and other cats in the home, although aggression towards people was lower if at least three cats were living in the home. One factor is easy for people to control: severe aggression was most common when people used positive punishment with their cat. See chapter 5 for more on training, and use positive reinforcement to train your cat.

Cat bites and scratches are most common to the hands and arms, but children are more at risk of bites to other parts of the body. According to one study, the most common circumstance for a bite is when the cat is afraid and trying to defend themselves from someone who is approaching or trying to pick them up.[23] Petting the cat and playing with the cat were identified as the most common situation in another study, where risk factors included being sensitive to handling and higher levels of general background stress in the home.[24] Cats bite the hands they live with; aggression towards people they don't know is much less common.

When cats are afraid of loud noises such as fireworks, they most commonly hide or try to escape, according to a survey of pet guardians published in the *New Zealand Veterinary Journal*. Shivering, trembling, and cowering were also sometimes seen at these times.[25] Always keep your cat indoors when it's likely there will be fireworks.

If your cat is fearful, don't force them to face their fears. This practice, known as flooding, risks making things much worse and may cause a state of learned helplessness in which the cat is too afraid to really do anything. Instead, try to think about things from your cat's point of view. For many fears, training can help, such as training a cat to like their cat carrier so that they will be less afraid of going to the vet. It may be possible to do desensitization and counter-conditioning training to help them learn to like the thing they are afraid of. If your cat is fearful of visitors to your home, ensure your cat has plenty of places to hide and always has an escape route. As well, don't let visitors approach your cat if the cat is fearful; instead, make visitors wait to see if the cat wants to go to them. Instead of having visitors feed treats, you can feed the cat while the visitor is in the room and then the cat does not have to approach the person they are afraid of to get the treat. For a very fearful cat, you may want to seek behavioral help and see your veterinarian to discuss medications.

In cases where you or others in your home are at serious risk from a cat's aggression, prepare a room that has everything they need in it and then isolate the cat in that room. They may follow a wand toy or food into the room, or if they are very aggressive you may prefer to use a sheet or towel to gently herd them. Then seek help with your cat's behavior. If you are bitten by a cat, clean the bite and seek medical attention. Cat bites can easily become infected because they penetrate the skin a long way, making a fertile breeding ground for bacteria. Children, seniors, and any-one who has a compromised immune system are particularly at risk of infection. A tetanus booster or rabies treatment may also be needed.

SEPARATION-RELATED ISSUES

IF CATS CAN become attached to their people, then it's possible that some cats might develop separation-related issues. However, very little research has been done into this issue in cats compared with dogs. Peeing and pooping in the wrong places while the person is out, being destructive, vocalizing excessively, and overgrooming are all signs, according to *Journal of the American Veterinary Medical Association*.[26] One survey of cat owners in Brazil, reported in PLOS ONE, found several behaviors that were described as occurring when the person was out of the house or out of sight of their cat.[27] They defined these as separation-related issues if the cat peed or pooped outside the litter box while the guardian was out, vocalized (e.g., meowed) excessively, or was destructive while the owner was out. As well, the researchers looked for emotional states of depression, aggression, agitation, or anxiety. They identified signs of separation-related issues in around 13 percent of cats, although this sample may not be representative of pet cats as a whole. The study found that these separation-related issues were more common when the cat did not have any toys and when there were no other pets in the house. Separation-related issues were also more common in cats who were left alone for longer periods of time.

Knowing whether a cat's behavioral issue is genuinely related to separation from their owner can be hard to tell. First, address and rule out any behaviors that might be caused by medical issues, boredom, or an inadequate environment, and see if you can get video of the cat when you're out. Your vet can advise you on medication, if needed. And you can use a gradual desensitization program that involves never leaving the cat for longer than they are happy with to train the cat to be content spending more time alone.[28] You may need the help of friends, family, or a pet sitter to achieve this. If you suspect your cat has separation-related issues, seek help from a veterinarian, veterinary behaviorist, or suitably qualified animal behaviorist.

ACTIVITY AT NIGHT

SOME CATS ARE awake and lively in the very early morning or even the middle of the night, disturbing their guardian's sleep. Medical conditions (such as feline cognitive dysfunction) can cause the cat to meow at night, so if your cat is waking you up for no apparent reason or shows other signs of illness (e.g., doesn't look well or is behaving differently in other ways too), it's important to see your veterinarian. But assuming medical issues aren't the cause, often it's a very mundane problem: the cat is bored and doesn't have enough to do in the daytime. And getting up to give the cat some food to shut them up can turn this midnight meowing behavior into a habit. Instead, give your cat more enrichment to ensure they have things to do, and in particular make time to play with them (e.g., with a wand toy or laser light) before bed. Finish the play session by giving them a little treat or snack. Since cats like routine, stick to the same time every night. As well, if they are howling at you for food, it may help to make the first meal of the morning come from an automatic feeding bowl that opens at a set time, so they don't come to you for the food. You can also leave out a food-puzzle toy for them to find and eat overnight.

"Listen harder when they are trying to 'tell' us something. Cats are notorious for hiding their emotions from humans, even their closest companions. To some degree this can be attributed to the fact that they evolved from small wildcats who are predators but are also prey animals. Prey animals often have adaptations which enable them to hide pain and fear. As a result, they are often a bit of an enigma to us. But as it turns out, they communicate a lot of information; it's just subtle. The angle of their ears, the movement or angle of their tails, or even the tone of their purr might communicate information about how they're feeling. So, if we listen really hard, and respect what cats are 'saying,' we can more accurately interpret how they're feeling and provide better care."

—**MALINI SUCHAK,** PhD, assistant professor, Animal Behavior, Ecology, and Conservation program and Anthrozoology graduate program, Canisius College

SEEKING HELP
....

ANY TIME YOUR cat has a new behavior issue, see your vet to rule out a medical problem that needs treatment. If the issue is confirmed as behavioral, you may want to seek help from a veterinary behaviorist (a veterinarian who is also board-certified in behavior) or a cat behaviorist (always ask about their certifications to ensure you are hiring someone qualified). Unfortunately, most people do not seek help for their cat's behavior issue, which means their cat continues to have issues and is potentially suffering from a lot of stress.

Several psychoactive medications (such as fluoxetine, more commonly known as Prozac) may be prescribed for cats with behavior issues. One survey found that most people would be open to using such medication if they thought it would help their cat, and they were more likely to say they would consider it if they themselves have experience taking psychoactive medication.[29] This finding shows that vets have an important role in explaining to their clients when such medication might be helpful, but also that people should simply ask their vet if they have questions about its use for a behavior problem; your vet may be able to alleviate your concerns. According to the first author of the study, Dr. Emma Grigg of the University of California, Davis, one of the most important findings is that "so few cat owners seek help for behavior problems, despite 97.8 percent reporting that they had experienced behavior problems with their cat(s)."

Veterinary behaviorist Dr. Karen van Haaften is also one of the coauthors of that study. She said, "Owner-perceived problem behaviors in cats are common, and most cat owners (93.5%) believed problem behaviors can be based in anxiety/emotional problems. Despite this, nearly half (49.8%) were unaware that psychoactive medications were an option for treatment of behavior conditions in cats, and many had concerns about potential side effects, especially sedation and potential for addiction. Veterinary professionals should be aware that cat owners come to conversations about psychoactive medications with pre-conceptions, and take the time to address these concerns and educate clients about risks and benefits of psychoactive medications. Cat owners reported the two most important decision-making factors were proven effectiveness and ease of administration."

When psychoactive medications are used with cats, it is often off-label, as they are labeled as being for dogs. More research is needed on the effectiveness of such treatments specifically with cats. I asked Dr. van Haaften to talk more broadly about when psychoactive medications should be considered as an option. "I mean the first thing I'd say is I don't think it should be considered a last resort," she said. "I do think it's one of the options and plan A or plan B often include medication for me, depending on the diagnosis." Before a veterinarian prescribes psychoactive medications, they must have made a diagnosis and a treatment plan. As well, she says, the likely prognosis with or without medication is also part of the decision. In addition, they take the welfare of the cat into account; for example, if the cat's welfare is poor, or if the cat's anxiety is so bad that it is a welfare concern in itself, then psychoactive medication may be considered.

"In cats I would say an example where I almost always try to use meds would be chronic urine marking where you can't get rid of the trigger," she said. "It's usually social stress from other cats, but if somebody lives in a neighborhood where there are lots of feral cats, and there's nothing you can do about the feral cats in your neighborhood, and it's stressing your cat out, and that cat is manifesting that stress by urine marking in your home. If you can't get rid of that trigger, like you can't move house or do something to get rid of the cats in your neighborhood, medication is an option and it's over 90 percent successful in fixing the problem. And there are very few side effects of long-term medication for that problem. So I'm going to recommend meds earlier rather than later because it's going to be a lot less stressful in the long run for that cat and for the owner. Who wants to clean up urine every day? That's a really quick way to break your bond with your cat."

In this study, people who had taken psychoactive medication also said they were more willing to try pheromones or cannabidiol (CBD) oils, but not herbal supplements. Unfortunately, the authors say the evidence to support the use of alternative approaches is "minimal." Regarding the use of CBD with cats, Dr. Adrian Walton told me, "The simple fact is we do not know. Cats, being obligate carnivores, do not eat much plant material, so their livers are not adjusted to dealing with toxins that plants produce to fend off predators, be they large or microscopic. Plants do not have fangs or teeth, but they are little chemical warfare systems that are not making these for our benefit. Just that in small doses can be helpful. So the answer is, I am less comfortable with CBD in cats than in dogs."

Making environmental changes is often an important part of resolving a behavior issue, because it reduces the cat's stress levels. Just like for us, stress affects the cat's whole body system, including their mental health, and can prevent them from being able to take advantage of opportunities for positive things. Stress can change a cat's behavior in many ways, according to *Journal of Feline Medicine and Surgery*.[30] It can increase meowing and other vocalizations, lead cats to hide more and to be more vigilant when out, and increase urine spraying as well as aggression and compulsive behaviors. At the same time, stressed cats may change their eating and grooming behaviors; facial mark their territory less; interact less often with their owner (and any other animals who live in the home) or alternately become clingy; play and explore their environment less; and generally be less active. As well, stress has been linked to feline interstitial cystitis (which results in toileting outside the box), an increased likelihood of upper respiratory disorders, gastrointestinal disorders, and dermatological issues.

When stress is contributing to behavior issues, it is important to work not only on removing the trigger (or changing how the cat feels about it) but also on decreasing overall levels of stress, because stress is cumulative (known as trigger stacking by dog trainers). That's why reducing stress by ensuring the cat has a safe place and by using other recommendations from the five pillars of a healthy feline environment (see chapter 3) and enrichment (see chapter 7) is so important. Pheromones may also be useful. When you want to change the cat's response to something, counter-conditioning is more powerful than desensitization (see chapter 5). Ensure that no one ever yells at, makes loud noises at, or squirts water at the cat, and try to give the cat control and choice as much as possible. As well, as cats get older, stressful things are more likely to bother them, which is why the next chapter looks at taking care of senior cats and those with special needs.

APPLY THE SCIENCE AT HOME

- If your cat develops a behavior problem, take them to the vet first to find out if there is a medical issue. If not, revisit the five pillars of a healthy feline environment and see what you can do to improve your home from your cat's perspective (see chapter 3).

- Never punish your cat for behaviors you don't like. Stress is often a contributing factor in cat behavior issues, and punishment will only increase stress and may affect the cat's bond with you.

- Check to ensure that the cat's needs are being met, because many behavior issues arise as a result of needs not being met.

- If seeking behavior advice for your cat, be sure to go to a suitably qualified professional. Don't be afraid to ask about their training and certification; they will be happy to tell you about it. Options include a board-certified veterinary behaviorist or a certified applied animal behaviorist (on referral from your vet); a certified cat behavior consultant; or, in the UK, a registered clinical animal behaviourist or an international feline behaviourist.

12

SENIOR CATS AND CATS WITH SPECIAL NEEDS

....

THE OTHER DAY when my husband and I were having dinner, Melina came to sit on one of the other dining chairs, as she often does. She likes to keep us company while we eat, and while I'm sure she's sometimes hoping for a morsel from our plates, mostly we eat food that cats don't like, like lentil stew or other vegetarian dishes. As I watched her, I noticed for the first time that one of her whiskers had turned completely white. It was a longer whisker and looked out of place compared with her other ones, which match her dark coloring. But a week later, the matching whisker on the other side of her face is also turning white. And now that I look closely, a few white hairs have appeared on her chest too. My Melina cat is showing signs of age, even though she is still a blur when she chooses to run round the house. And Harley, who is a year older, also has more gray hairs.

Cats go through a number of changes as they age. According to a survey on American cats, around 20 percent of them are seniors. That's a lot of older cats that may need a little extra help to keep them comfortable as they age. And there are a number of conditions that may affect cats at any age. Let's start by looking at senior cats and then consider some of the other ways we can help cats with special needs.

HELPING CATS AGE

CAT YEARS DON'T map exactly to human years, particularly since so much growing up is done early on. Kittens grow up fast, and pretty soon your young cat is in their prime (3–6 years). Between 7 and 10 years, your cat is mature or middle-aged, according to the American Association of Feline Practitioners.[1] Once over 10, they are seniors, but it depends on the individual; some may be senior by 8 years. Cats who are well cared for (and lucky too) can live into their late teens or even early twenties. These numbers are generalizations, and so the individual cat in your home may age earlier or later than the norm.

Not all aging is bad. Some is neutral, as with Melina's new white whiskers. She isn't about to look in the mirror and have a midlife crisis about it, and it's just a cosmetic change. Some aging is even a good thing: your older cat knows your routine and has some pretty fixed habits, and hopefully you are all settled into a happy coexistence. They aren't likely to chew your wires, climb the Christmas tree, or curl up on a warm stove top like a kitten would do. They are wiser to the ways of the world, and you have had time to train them. Some aging is definitely not good, however, and as cats age they are more prone to particular diseases.

Aging affects all of the body's systems, and a lot of these normal changes are similar across cats, dogs, and humans. As your cat gets older, they may begin to lose some of their senses, as their vision, hearing, or smell becomes impaired.[2] They may have issues to do with their heart or circulation, and their lung capacity will be lower than when they were younger. You may notice changes in how they move, such as a reluctance to jump as high or stiffness due to issues such as arthritis.

Changes often occur in a cat's skin and coat as they age. They may develop more white hair, and although two of Melina's whiskers have turned white, paradoxically one or more whiskers may turn black as cats get older. White hairs are due to a loss of the cells that contain pigment (melanocytes) due to age. As well, with age an enzyme called tyrosinase is less active; this enzyme contains copper and is involved in melanin production, so less activity means more white or gray hairs. Older cats will also have less skin elasticity, and their nails are more brittle than when they were younger. They may have sun damage if they have spent a lot of time outdoors, and skin cancers are more common at this age.

Many changes are related to normal aging but you shouldn't just assume a new condition is due to age. If you notice any changes in your cat, always take them to the vet. According to a study on senior cats published in *Veterinary Sciences*, the most common behavior changes in cats aged 11 and older are being more affectionate towards their guardians, vocalizing more (including at night), and house soiling.[3] Other behavior changes include eating less, drinking more, grooming less, playing less often, going outside less, and sleeping more. The researchers point out that many of these behaviors may have a medical cause. Common diagnoses in these senior cats were dental issues, an overactive thyroid, urinary tract disorders, and kidney disease.

Dr. Patrizia Piotti is a veterinary behaviorist and a postdoctoral researcher at the Department of Veterinary Medicine, University of Milan, Italy. She says that we need to distinguish between healthy and unhealthy aging. "The message we are trying to pass on as researchers and vets as well," she said, "is that if your cat has reached a certain age and is not interested in the environment, sleeps all the time, and has a number of issues which might be house soiling—or even seems in pain or struggles with eating, for example, then you shouldn't automatically say well this is normal because my cat is old. It's really important to look into what might be causing these problems." She says that you should take your cat for regular vet exams because some of the conditions that tend to affect older cats—an overactive thyroid, high blood pressure, and kidney disease—can start to appear in cats at a younger age, even 7 or 10. These conditions can be easily managed for many years, she said, so it is important to find them before you see signs.

Dr. Piotti says to also look out for symptoms of cognitive dysfunction, which are known by the acronym DISHAA:

Disorientation (such as getting lost in the house or stuck somewhere)

Interactions (changes in how the cat greets or interacts with people)

Sleep-wake cycle changes, such as being awake at night

House soiling

Activity-level changes

Anxiety

However, she says, "cognitive dysfunction syndrome is a diagnosis of exclusion, so you can only diagnose it after excluding all other conditions." In other words, you need to rule out a range of medical conditions as well as behavioral conditions such as stress, so talk to your veterinarian if you see changes in behavior, including the cat no longer behaving in ways they used to.

Older cats need to have easy access to their resources and good places to sleep, given they spend more time sleeping. Steps or a ramp can make it easier for your cat to get up to a nice perch or onto your bed. If they are thinner, they may especially appreciate more padded bedding to help them feel comfortable. They may not groom themselves as much, or may not pay as much attention to areas like the tummy and trousers that can be harder to reach due to joint pain. They also produce less sebum, an oily, waxy substance that helps to keep their fur looking good. So you may see changes in your cat's coat. Even if they never needed to be brushed before, you may need to brush them as they get older.

Grooming your cat is a good way to keep an eye on their weight and body shape. Some cats become overweight or obese because they are less active, whereas others start to lose weight due to medical issues. Weight loss is quite common in older cats and can be due to many causes, including diabetes, hyperthyroidism, and intestinal diseases. As well as losing weight, from the age of 11 cats may lose lean body mass; just like people, they have less muscle as they age. Gradual changes in weight can be difficult to notice, so it's a good idea to weigh your cat regularly (or get your vet to). Your vet will be looking not just at overall weight but also body condition. If your cat is less active, they may need less calorie intake, but pay attention to the quality of the food.

If your cat has more than one disease, it may be tricky choosing a diet for them.[4] For example, a therapeutic diet may have too many calories, too much protein, or inadequate amounts of other nutritional elements. Pay attention to what you see in the litter box when you scoop it and to what your cat is eating, and discuss any changes with your vet.

Cats can be picky eaters at any age, but it's more common for senior cats to be finicky about their food. Medical issues such as chronic kidney disease, dental problems, or memory issues can make cats less likely to eat or less likely to eat as well as they used to. So if your cat is a finicky eater, first check with your vet. Then, try different options to interest your cat in eating. For example, offer a range of different foods for them to choose from. Change the texture of the food, such as from a chunky food to a pâté, a food with smaller pieces in it, or a food with gravy. Sometimes you may need to add a small amount of water to the food. Warming it in the microwave gives off an aroma, which makes it more palatable. You can also add a small amount of unseasoned broth (make sure it is low in sodium and does not contain garlic or onion) or the liquid from a can of tuna.

Canned food is generally recommended for senior cats to ensure they are getting enough water, but for cats who prefer kibble, adding water or broth may be another option. Ensure that senior cats have access to a water fountain or ice cubes if they like those. If you are changing your cat from one food to another, do it gradually by offering portions of both foods and increasing the amount of the new food as you decrease the amount of the old food.

"People should accept cats as individuals, with rich personalities and complex social needs. People who haven't had a close relationship with a cat often assume the stereotype that cats are 'independent' and just 'use' people for food and warmth. This leads some people to consider cats as an option for a pet that requires less attention and responsibility than a dog. However, as I'm sure cat lovers will be happy to tell you, cat personalities differ greatly, as do their social needs. Many cats are capable of great affection, and if given a choice would choose to be in your company rather than alone, which can mean their welfare is easily compromised when left alone for long periods or are shut outside all day."

—**NAOMI HARVEY,** PhD, Canine Behaviour and Research Team, Dogs Trust

CATS WITH SPECIAL NEEDS

CATS CAN HAVE a range of special needs, including syndromes they may be born with or issues that develop later. Luckily, cats with special needs can still have a great quality of life. You may need to make a few accommodations to assist them and to help them feel safe.

Vision loss

When pets lose their vision, the most common concern of their owners is for the pet's quality of life. Other common concerns are whether the pet will become depressed and anxious and how well they will adapt to the change, according to a survey of veterinary ophthalmologists in *Veterinary Record*.[5] The good news is that cats typically adjust well to losing their sight, and there are many things you can do to help. The three most useful things are keeping the home environment the same, avoiding making unnecessary changes to routines, and using your voice to help the cat.

Keeping the environment the same means that your cat knows where all of the furniture is and won't bump into it. As well, they know where their stuff is and can always find it. If your home has any dangerous areas, you will want to restrict access by closing a door or installing a pet gate. Think carefully about how your cat is going to cope with heights in your home. For example, make sure they cannot accidentally fall from any large drops such as from a landing, and consider how to make the stairs safe or if they should be blocked off. If your cat has been allowed outside, help them adjust to being indoors or only allow access to an enclosed patio to keep them safe.

A routine is a good idea for any cat, and especially for a cat with special needs. If they are adapting to vision loss, try hard to keep to the same routine they have always had, and make sure they still have the same opportunities to interact with you. Scent can be an especially good way to provide enrichment for cats with special needs; you may like to revisit some of the suggestions from chapter 7. If your cat seems disoriented, use scent to mark reference points for them. Put a scented item in a certain location and leave it there as a fixed reference point. But remember, avoid strong scents the cat will not like! You can also ensure that the cat has different tactile surfaces to experience—soft, plush fur on a cat bed, for example, versus crinkly paper for a cat tunnel.

Even if your cat can't see well, their hearing may be fine. Use your voice to guide them. For example, call them over to you or alert them that they are about to bump into something. Give other pets in the home a bell on their collar so that your cat can hear them coming. Ensure that cat toys have a scent and/or noise associated with them; you may still be able to engage them to chase a toy that has a little bell or that squeaks.

In some cases, you may also want to discuss pain medication with your vet. Glaucoma is a common cause of blindness, and in people it does affect quality of life. The study in *Veterinary Record* says that glaucoma in animals can be painful, and so your veterinary ophthalmologist should discuss pain management with you for this condition.

Hearing loss

Cats have very sensitive hearing and can hear sounds from 45 hertz to 65 kilohertz, which is both lower and higher than humans can detect. Being able to hear the high-frequency (ultrasonic) sounds from mice is useful in hunting. Hearing impairment or loss can therefore affect cats' ability to hunt and also to hear predators or vehicles.[6]

Some cats are born deaf. Inherited congenital deafness, or deafness from birth, is common amongst white cats with blue eyes (although not all white cats are deaf, and if they are, it may not affect both ears to the same extent). A gene suppresses melanocytes in the fur, making the fur white, and in the eyes, making the eyes blue, and it also suppresses melanocytes in the stria vascularis of the ear. The melanocytes in the ear help to maintain high potassium levels; without them, there is damage to hair cells, causing deafness within a few weeks of the kitten's birth.

Deafness can also be caused by other genetic factors, certain drugs, exposure to loud noises, infections, and ear mites, and of course aging. Although some types of deafness cannot be treated (such as congenital deafness and old-age hearing loss), others can, so see your vet if you think your cat is deaf. A cat that does not respond when you call their name or shake their favorite treat packet, does not greet you when you come home (because they didn't hear you come home), startles if you approach them from behind, and no longer responds to noises they don't like (e.g., if they normally jump when the coffee grinder is turned on) may be experiencing hearing loss or deafness.

A basic behavioral test for hearing impairment involves making a sound out of the cat's sight and seeing if they startle or move their ears. This test is not reliable, as cats may still detect the movement that made the sound (e.g., from air currents). As well, cats who can hear may be too stressed to respond, or may be uninterested. If you try this test at home, pick sounds that cover a wide range of frequencies, and make very soft noises to start with. Increase the volume only if you get no reaction. The brainstem auditory evoked response (BAER) test is more reliable and can be done in clinics. Your veterinarian will be able to help you get your cat's hearing tested.

Cats with impaired hearing are best kept indoors so they are not at risk of being killed by a predator or hit by a car. Try not to startle the cat, and in particular, ensure that toddlers and small children do not do so. A startled cat may scratch or bite. They may be able to detect vibrations in the floor that will alert them to your approach. If you want to train them, you could use hand signals, which are often used in the early stages of training anyway. For example, you could teach a cat to come when they see light from a flashlight by showing the light and then giving a treat. Start when the cat is close by so they do not have to come for the treat, and then slowly increase the distance. Always give a treat when the flashlight is on, or you will begin to extinguish the response (see chapter 5).

Mobility issues

Kittens can be born with cerebellar hypoplasia, or they can develop it when a queen passes on feline parvovirus (FPV) or when they catch FPV during the first weeks of their life.[7] It is a condition in which a part of the brain called the cerebellum is not fully developed. Kittens with cerebellar hypoplasia have poor balance, are wobbly, and may fall down often. They may have a funny gait that is like a goose step and an intention tremor that occurs when they are thinking about movement. The signs don't become apparent until the kitten starts to move around. Sometimes an MRI may confirm that the kitten's cerebellum is not fully developed.

Cerebellar hypoplasia has no treatment, but kittens with this condition will go on to live happy lives. It is a good idea to keep cats with cerebellar hypoplasia indoors, as being wobbly puts them at greater risk in their environment, including from predators. As well, you may need to block off any heights in your home (e.g., from a landing) so they cannot fall, and look out for other ways in which they may need a little extra care.

Cats who have had a limb (or their tail) amputated also have mobility issues. According to a study of more than 200 cats who had had an amputation, young males who have been in road traffic accidents are the most common amputees.[8] Other reasons for amputation include damage to the nerves or to the skin and muscles. Most often, a back leg needs to be amputated (the front legs bear most weight for cats). Almost all cats were given pain relief following the surgery, but some guardians said their cat still experienced pain after returning home, suggesting additional pain relief might have been needed. The good news is that 89 percent of these cats were said to have a normal quality of life. To help tripod cats adjust, you may need to move furniture around and add steps (to help them reach places) or modify the litter box so they can continue to use it without difficulty.

If you have concerns about how these or any other conditions may affect your pet's quality of life, speak to your vet. As well, there are good forums and Facebook groups that can provide support to guardians of pets with various conditions. Having said that, always question the quality of the sources you find on the internet, because there is a lot of misinformation out there.

APPLY THE SCIENCE AT HOME

- Keep an eye on your cat as they get older. Don't just put changes down to old age—see your veterinarian in case the changes are signs of a condition that you can treat.

- If your cat becomes a finicky eater, check with your vet in case there is a dental or other medical cause. Try different foods, softer foods, adding water, and warming the food to encourage them to eat, and ensure they are getting enough water.

- Try to ensure your cat's life is not stressful and always ensure they have an accessible hiding place if they want to retreat and be on their own. Older animals don't tolerate stress as well as younger ones.

- Be ready to make some adjustments to your home to help a senior cat or a cat with special needs. For example, if they are no longer able to jump high up like they used to, move things to a lower level or provide a ramp.

13

THE END OF LIFE

· · · · · · · · · · · · · · · · ·

LOSING A PET is the hardest part of being a pet guardian. I lost my Australian Shepherd, Bodger, to cancer in the early stages of writing this book. Grief can be a hard thing to manage. These days, we know that the loss of a pet is a real loss, and fewer people say that "it's just a pet." Because we so often have to make decisions about our pet's end of life, it can be helpful to think ahead about what is important to you and what you would prefer to do when that difficult moment comes. But if you have lost a pet recently or find this topic difficult, it's fine to skip this chapter for now and come back to it when you are feeling more resilient. The good news is that cats can be quite long-lived pets.

HOW LONG DO CATS LIVE?

ALTHOUGH THE AVERAGE age that a pet cat lives to is between 12 and 15 years, cats can live much longer, and a handful are known to have lived up to age 30.[1] One epidemiological study in the UK of a random sample of just over 4,000 cats found the average longevity was 14 years.[2] In fact, the data showed a bimodal distribution, with two main peaks. One peak showed that some cats died young, at 1 year old, and the other peak in the data was for a life span of around 16 years of age. Mixed-breed cats lived longer, with an average longevity of 14 years, compared with purebred cats who typically lived for 12.5 years.

In dogs, mixed breeds generally live longer than purebred dogs because of "hybrid vigor" (although it does depend on the breed). It is thought that breeding for certain traits of appearance may negatively affect the health and longevity of the breed. It seems likely that a similar process is at work for pedigree cats, and that greater genetic diversity builds resilience. However, there was a lot of variation in the data for pedigree cats, which suggests that some breeds are affected more than others. In this study, the breeds with the shortest life spans were the Bengal cat (7.3 years) and Abyssinian (10 years), although it is important to note that the data for pedigrees is based on only a small number of cats. Ragdoll, Maine Coon, and British Shorthair cats also had shorter-than-average life spans. The cat breeds with the longest life span were the Birman (16.1 years) and the Burmese (14.3 years), while Siamese and Persian cats had life spans similar to those of mixed-breed cats.

The most common causes of death for the cats in the study were trauma (especially road traffic accidents but also animal attacks) at 12 percent, kidney disorders (12%), illness (unspecified, 11%), cancer (11%), and mass lesions (which likely included many cancers, but the exact diagnosis was not known, 10%). For young cats, the most common causes of death were trauma (47%), viruses, and respiratory disorders. Trauma was much less common in older cats, where cancer was a more common cause of death. The study took place in the UK where most cats have outdoor access for at least part of the day.

Cats tended to live longer if they were mixed breed, had a lower weight, had been spayed or neutered, and, interestingly, were not insured. It seems possible that pedigree cats are more likely to be insured, and so it could be that not being insured overlaps with being a moggy (mixed breed). But the study looked only at whether or not the cat was insured when it passed, and it is also possible that people with healthy cats may have canceled an insurance policy at some point, as they may not have thought it was necessary.

These results show the importance of regular checkups at the vet, and, in particular, checkups to detect signs of kidney disease early on, as that was a common cause of death. A higher body weight was associated with a reduced life span, showing the importance of monitoring your cat's weight and avoiding overweight and obesity. The study found that for cats who lived to at least 5 years of age, those who weighed between 4 and 5 kg (8.8 and 11 pounds) lived for 1.7 years less, on average, than those who weighed less than 3 kg (6.6 pounds).

A study of the causes of death for cats in Taiwan found similar results.[3] The most common causes of death were kidney and urologic disorders, cancer, infections, issues affecting multiple organs, cardiovascular disease, and trauma. The most common infection was with feline coronavirus. Many cats get feline coronavirus (up to 40%), and it typically causes only mild disease, although sometimes there are no symptoms at all. However, the virus occasionally mutates as it replicates within the cat and becomes feline infectious peritonitis (FIP). FIP is most common in young cats and is almost always fatal.

Of course, for many cats, the death is due to euthanasia. In this study, over 85 percent of the cats were euthanized. This can be a very difficult decision for the cat owner to make, and one factor that plays a large role in the decision is the cat's quality of life.

QUALITY OF LIFE AND EUTHANASIA

THERE'S A COMMONLY held belief that pet guardians will "know when it is time" for their animal's euthanasia, but this assumption glosses over the emotional difficulties of making such a decision and the very different experiences people may have had getting there. Ethical considerations, the human-animal bond, and the circumstances all play a role.

From an ethical perspective, the "right" time is when the cat no longer has quality of life but before they begin to suffer. But that time is hard to gauge. There is a risk of making the decision to euthanize a pet too soon, when they still have some good days, weeks, or months left, and, conversely, of waiting too long, by which time the pet is suffering.

The human-animal bond may also play a role in this decision.[4] In some cases, the strength of the bond with the pet may cause people to want to delay euthanasia, because they cannot bear to lose their pet, potentially prolonging the animal's suffering. However, a strong bond may also cause some pet guardians to elect for an early euthanasia, sooner than is necessary, because they don't want to see the animal suffer.

Having to consider euthanasia when your cat has had a traumatic injury (such as after a road traffic accident or attack by another animal) is different from making that decision when they are ill, such as with cancer, and their quality of life changes day by day, including some ups and downs, over a long time. Knowing whether one bad day is the day to make the decision, or is no different from the other bad days that were followed by better days, can be difficult.

Your veterinarian's opinion can often help. Veterinarian Dr. Adrian Walton of Dewdney Animal Hospital in Maple Ridge, BC, says that both the pet's and the owner's quality of life are important parts of the decision-making. "One of the things that I say as a veterinarian is my job is not to save the life of your pet; my job is to save your relationship with your pet," he says. He adds, "One of the things I'd be saying is if you're finding that this is negatively impacting your life, then euthanasia's on the table. But as for the pet, the simplest way is, when the bad days outnumber the good then certainly euthanasia's on the table."

Several quality-of-life scales exist to help people make decisions about their pet's quality of life. Most of these are designed for specific conditions, such as cancer or arthritis, and to be used with a vet's involvement. It's important to know that quality-of-life scales are not designed purely for euthanasia decisions, but they may help identify proactive steps you can take to prolong the time your kitty has a good quality of life. Medication, changes to the environment, or changes in routine are all possibilities.

A lot of people seem to equate good quality of life with eating, or with playing, but of course it is important to pay attention to the full picture. For example, a cat who is not eating may have dental issues or sores in their mouth. If you can alleviate those conditions, the cat may start eating again and have a good quality of life for some time to come. Remember that other behaviors also represent a positive quality of life, including having a good appetite, purring, kneading the paws, looking for attention, seeming happy, being affectionate, and being interested in things.

MAKING END-OF-LIFE DECISIONS AND WHAT TO EXPECT

DR. KAT LITTLEWOOD is a veterinarian in New Zealand and a lecturer at Massey University, where one of her studies looks at how owners decide to euthanize their cat and how vets can best support them. She spoke to fourteen people who had euthanized their senior cat, as well as to their vet, to find out more about the process. The cats discussed in the study were very old. She says her participants included "a lot of cat owners that were very, very attached to their cats, so they found it really difficult. Particularly when it was a slow progression, and their cat went downhill quite slowly, they found it really hard to make that call. And because cats live so long as well, a lot of the cats in my studies were 19 years old. That's a long time to be with someone." At the time of Littlewood's interviews, New Zealand was conducting a referendum about whether or not euthanasia should be allowed for people—termed medical assistance in dying (MAID) in North America—and she found that this subject kept coming up in what people said. It's perhaps not surprising that people's wider beliefs about death and dying will influence how they feel about the end of their cat's life.

I asked Dr. Littlewood what advice she had for people making end-of-life decisions about their cat. One thing she said was to think about things that are important for the individual cat. For example, some cats may like being outdoors, so not going outside could be an important sign for the owner. For other cats who are not so keen on going outside, it may not matter very much if they are choosing to stay indoors. She found that people were willing to help their cat eat and drink, such as finding tastier foods or hand-feeding the cat, but felt less sure of what to do about changes in their cat's interactions with them. She says, "Because a lot of the cats were older and so there was a little bit of difficulty that the owners had distinguishing between changes that resemble an aging cat versus changes that are like, hmm, it's not really so great. So being aware of where that cat's heading and what's normal on a weekly or even monthly basis [can help]. Maybe having a little check-in with things like how much is my cat eating, what is my cat doing. Just having a little checklist of things and checking it over to see how that's doing would probably be ideal."

Unfortunately, however hard the decision to euthanize, most cat owners must make it at some point. Dr. Walton has some words of caution for people who think they would prefer to let their pet die at home without assistance. Especially when cats have kidney disease, he says, this can go horribly wrong:

"When they're 15 years old or older, those cats will die of kidney failure," he says. "Ninety percent of them. Kidney failure is not like other forms of death. Heart failure, when they start to go downhill they go down quick. Kidneys are different. These are desert animals, these guys. Urine's been building up in their system for a long time and what eventually happens is, they'll stop drinking. When they stop drinking, people think of it in human terms, 'Oh, you can only live three days without water.' So they'll say, 'Okay, we're just going to have the cat die at home.'

"So three days go by, and the cat's still sitting there in his bed not doing anything. And they go, 'Well it can't be much longer' and another three days go by. Now it's been a week. Now they're saying, 'Well I can't go to the vet, he's going to think I'm horrible.'

So they wait another week. Now they're starting to panic. The cat's still there. He's looking like death warmed over but he's still breathing. He's not dead yet; he's just slowly dying. And this can take up to three, four weeks to die of dehydration. So if your cat's in kidney failure and stopped drinking, come in."

I asked Dr. Walton to describe the euthanasia process to me, and he said what surprises many pet owners is how quick it is. People expect their pet to disappear "over a period of five minutes, when actually it's more like thirty seconds." One thing that can surprise people, he said, is that the eyes stay open after death. And people often aren't aware that some bodily processes happen. One is agonal breathing, a reflex that occurs when carbon monoxide builds up in the bloodstream after the heart stops, causing the chest to expand. The body may also twitch or stretch, and the bladder or bowels may release. And if your pet does die overnight at home, then rigor mortis may set in, which means the body becomes stiff.

As well as thinking about when the moment comes, it is also worth thinking about how and where it should happen. Dr. Littlewood told me, "Something owners need to be aware of is that they have more power than they believe. Some of my owners didn't realize that they could request a certain vet to be with them. Owners didn't realize that there was the opportunity to do a home euthanasia. So I think that some advice I would have for people, cats or dogs for euthanasia, is to think about things—and actually thinking about what makes you and your cat the most comfortable. Because oftentimes the vet can come to your home for euthanasia, and it's probably going to be better for a lot of cats because a lot of cats aren't as adapted to being in carriers as we would like them to be."

"Sooner rather than later: If owners and their veterinarians considered death earlier, then quality of life would be better for cats. Most cat owners tell us, after they have euthanized their beloved cat, that they wish they had done it sooner. It can be really difficult to make the decision at the time—especially if owners are strongly attached to their cat and/or think of them as a member of the family. Owners also adapt to their cat's declining quality of life, and poor welfare becomes the new normal for their pet. However, for the cat's well-being, it is better to start thinking about 'how it is doing' sooner rather than later. I would like veterinarians and owners to work together to make judgments on quality of life—before it is significantly compromised. We need to break down the taboo of talking about death. When a cat is diagnosed with an illness, particularly if it's terminal, its death should be discussed. When a cat reaches a certain age, we need to have 'that' conversation. What does a good life look like for this cat? How will we know when it is no longer enjoying life? By having these frank discussions earlier, we can help reduce the 'wish I had done it sooner' effect and improve the quality of our cats' lives."

—DR. KAT LITTLEWOOD, BVSc (Dist), PGDip (Dist), AFHEA, MANZCVS (Animal Welfare), PhD, small-animal veterinarian and lecturer at Massey University

PREPARING FOR GRIEF AND LOSS

REMEMBER THAT GRIEF is a normal reaction to losing a pet. Your cat was a member of the family, and their loss is significant. Be gentle with yourself, and surround yourself with people who can support you. Unfortunately, some people will still say, "But it was just a cat." Thankfully, more and more people understand that this is a loss with depth that will take some time to get over.

Sometimes people can feel anticipatory grief before the loss itself, and find that even thinking about when it might happen makes them very anxious or changes how they interact with their pet. Dr. Naomi Harvey wrote movingly of how she realized she was treating her cat, Dreamer, differently because she was afraid of losing her.[5] Once she realized this was anticipatory grief, she was able to go back to giving Dreamer her usual amounts of play, fuss, and attention.

One study reported in *Anthrozoös* found that, not surprisingly, a strong attachment to the pet was linked to feelings of grief and sorrow on their loss, as well as anger (but not guilt).[6] What might be called "complicated grief"—a more difficult reaction to the loss of a pet that may include a longer-lasting depression—tends to include feelings of anger and/or guilt as well as grief and sadness. Anger was more common when the pet's death was sudden, and both anger and guilt were more common when the reason for the pet's euthanasia was some kind of cancer. Although it's understandable that a sudden death may lead to anger, it is perhaps a bit harder to interpret the result relating to cancer. Many factors may be at play, including that sometimes animals must be euthanized soon after a cancer diagnosis, or that treatment such as chemotherapy or radiation therapy is an expensive and stressful process for the owner to go through.

Memorializing your cat can help you cope with your grief. Many people like to keep a paw print, and some like to either keep or scatter the ashes after their pet is cremated. There are also websites where you can write an online memorial, and of course many of us like to share messages about and photos of our cat on social media or in emails with friends. You don't have to make decisions about whether to keep, throw out, or donate items that will remind you of your pet until you are ready. Lean on your friends and family at this difficult time, and take care of yourself.

HOW PETS COPE WITH THE DEATH OF A COMPANION

AFTER BODGER PASSED, Melina spent a few days seeming to look for him. Having been accustomed to sleeping in his bed during the daytime, she would not do that on the first day without him. She looked a bit lost, as if she was not sure what to do. The second day, she spent a little while wandering around looking unsure, and then decided she would use the bed after all. It seemed to me that she missed him. Harley was harder to read. He seemed to notice Bodger was missing, but his behavior did not change much and he did not seem to feel the loss as deeply as Melina.

Some pets do miss their animal companions when they pass away, but for some it's less obvious, at least according to a survey that asked both dog and cat owners how their pet responded to the loss of an animal companion.[7] Around three-quarters of cats (and around the same proportion of dogs) were said to show behavior changes after an animal companion passed. The most common change, in almost all of the affected cats, was to be more demanding of affection and more clingy generally. Other common changes in cats were to meow more often and more loudly, to show aggression to other cats in the household, and to seek out the places where the deceased animal used to spend time. Unlike dogs, cats' eating behavior was not reported to change. In some cases, the pet had seen the body of the animal who died (either because it died naturally at home or the euthanasia took place at home) but this did not have any effect on the behaviors described.

The behavior changes decreased over time, and all of the pets were said to be back to normal six months later. So if you lose one pet but still have other cats in the home, be kind to them and let them have more fuss and affection if they seem to need it.

PLANNING FOR THE UNEXPECTED

IT IS IMPORTANT to make plans for your cat(s) in case of emergency and in case you are hospitalized or predecease your cat. When thinking about who might take care of your cat if you are ill or pass away, consider not just who loves cats but also who has a similar approach to pet parenthood as you. If your cat has a good relationship with a friend or family member, they could be a good person to ask. Of course, the person who will help in case of temporary emergency does not have to be the same person who will take care of your cat if you pass and who might be a relative living somewhere else. It helps to have someone who is trustworthy, has the key to your home, and knows what to do about cat food and cat litter if you can't make it home when you expected to. Some local rescues and shelters offer emergency boarding and might be able to care for your cat in the short term if, for example, your house catches fire. When going on vacation, ensure that any pet sitter or cat hotel is a member of a professional body and insured.

Being able to evacuate with your pet is essential in case of natural disasters such as wildfires, earthquakes, hurricanes, and so on. Plan in advance for how this might work. Since some hotels do not accept pets, make a note of hotels near you that will allow pets to stay. We are all advised to have a grab-and-go kit in case of emergency, and this applies for your pet too. Things to include in your kit include copies of your pet's medication and vaccination records, phone numbers for your vet and emergency vet, enough food and bottled water to get through at least seventy-two hours (or potentially up to a week), food bowl, cat litter and litter box, some cleaning supplies and garbage bags, some bedding for your cat, a brush and a washcloth if they need regular grooming or their eyes and face need to be wiped, and a few toys to help them feel at home. You may also want to include a crate that your cat could live in temporarily if you had to go to an evacuation center. Of course, you will need to put your cat in their carrier if you have to evacuate, and this will be much easier if you have already trained your cat.

Your cat should have some sort of identification (a microchip or tattoo) in case they ever get lost. One study found that vets are less likely to recommend microchips to cat owners compared with dog owners,[8] so ask your vet, and remember that even indoor cats can escape to the outdoors. Remember to keep your address and phone number current with the microchip database (or with the vet for a tattoo). It is also a good idea to teach your cat to come when called, even if they are an indoors-only cat (see chapter 5).

Make sure you have a photo of your cat that shows clearly what they look like. If you ever need to make flyers, you'll have a photo to use. When cats do go missing, the sad news is that many cats who arrive at local animal controls and shelters are never reunited with their owner. While 15 percent of pet owners had lost a cat or dog in the previous five years, only 75 percent of cat owners were reunited with their cat, according to a study published in *Animals*.[9] A five-year study of what happened to cats arriving at the RSPCA in Queensland, Australia, from 2011 to 2016 found that only 5 percent were returned to their owners; in another study of councils in Victoria, Australia, 13 percent of cats were reclaimed by their owners.[10] To reduce the risk of your cat going missing, keep your property secure, fix any broken or flimsy screens, and make sure that all occupants and visitors know the rules about open doors and windows, etc.

When cats go missing, it is often after leaving via an open door or garage (74%), but some escape via a window (11%), through a broken window screen (6%), or from a balcony (5%), according to some other research.[11] A third of missing cats were found alive within seven days, and half after thirty days. By sixty-one days, only 56 percent of the cats had been found, and beyond this time only a few showed up. Indoors-only cats were found 39 meters (128 feet) from home, on average, and indoor-outdoor cats 300 meters (984 feet) from home (although this difference was not significant). For all the cats (indoors, indoor-outdoor, outdoors), the median distance from home was 50 meters (164 feet), and 75 percent of cats were found within 500 meters (1,640 feet). This suggests that you should look very carefully very close to home if your cat is missing.

Remember that cats can hide in tiny spaces and also like to be high up, so look at different levels too. Searching on foot is the most successful strategy, and it is also worth looking after dark. Use a flashlight to look behind books on the bookshelf, at the back of the closet, or inside a cupboard. Think like a cat about little places that might offer the opportunity to hide or get cozy and warm.

Other strategies to help you find a missing cat include putting up flyers, asking neighbors to check outbuildings and garages, making a hiding place with a cardboard box by your front door, and shaking a treat packet in case it entices the cat to come out (but remember, missing cats are often scared and may be too afraid to come). Listen for your cat meowing and consider setting up a trail cam in your yard. Some local rescues or shelters will lend you a live animal trap (typically for a small fee or donation), although you will have to monitor it and be willing to free any wildlife that ends up in there. If your cat gets stuck up a tree, call a local arborist who can go up and get them. And remember to keep searching and not give up hope. Of course, since a small proportion of people acquire their cat by finding it, if you do find a cat, always get them checked for a microchip and put up flyers (online and in physical locations such as local noticeboards) to try to find the owner.

APPLY THE SCIENCE AT HOME

- If your cat has a health condition, see if you can make adjustments or if your vet can recommend treatments that will help to prolong a good quality of life.

- Think in advance about how you will make end-of-life decisions. A good general rule of thumb is to consider euthanasia when the bad days outnumber the good. Remember to consider what is important to your cat. In some cases, when the time comes, vets will come to your home so that your cat does not have to go to the vet one last time.

- Make a plan for who you would like to take care of your cat if you get ill or predecease them, and update your plan from time to time or when circumstances change.

- Make sure your family's emergency plan includes your cat(s). If they aren't used to going in their carrier, start to train them. Find out which hotels near you are pet friendly. Make an emergency kit that includes enough supplies for a few days and ensure your pet has identification such as a microchip.

14

HOW TO HAVE A HAPPY CAT

. . . .

R IGHT NOW, BOTH cats are sleeping. It's a cold day in early
November, so Harley is on the heat vent absorbing as much
warm air as he can, and Melina is tucked up under our bed
in a blanket on what used to be Bodger's dog bed. I peeked under
there to have a look and saw her ears pop up and the orange stripe
on her nose become visible, and she looked at me like she didn't
want to come out of that nice, cozy space. So I left her be.

One of the things that are important for cats is agency: getting to make their own decisions. Other things cats like include
routine and predictability, but also play, nice food, petting in the
right places and for the right length of time, and so on. There are
the things that are important for all cats, and then the things that

matter for your cat. To make our home right for our cat(s), we need to provide the five pillars of a healthy feline environment (chapter 3). And for their welfare overall, we need to provide good nutrition, a good environment, good health, behavioral inter-actions that give opportunities to express normal behaviors, and opportunities to have positive experiences. And because positive experiences can't be enjoyed if your cat is feeling afraid or anxious or in pain, we need to keep negative experiences as low as possible.

As we've seen in this book, feline science is both fascinating and useful. It tells us a lot about what our pet cats need (see the table on page 232), but at the same time, much more research is needed in a variety of areas. Many studies are only small or on laboratory cats and may (or may not) apply to pet cats in homes. Some standard pieces of wisdom about cats turn out not to have much evidence behind them, although in many cases it makes sense to keep recommending them (like using food-puzzle toys or providing an extra litter box). It is wonderful to see so much interest in feline science these days, and it makes me happy to think that many more studies to come will give us an even better idea of what we can do to make our cats happy (or even happier).

Good welfare for pet cats

NUTRITION	PHYSICAL ENVIRONMENT	HEALTH	BEHAVIORAL INTERACTIONS
• Good food • Multiple small meals a day • Water	• Safely enclosed • Multiple, separated resources • Takes account of the cat's sense of smell • Nice cat beds • Good scratching posts • Safe places to hide or relax, including away from children • Appropriate bylaws and legal framework	• Good health • Low-stress veterinary care • Training for body handling and grooming • Breeding for good health and to avoid hereditary conditions • Exercise	• Positive experiences during the sensitive period for socialization • Companionship with or apart from other cats • Companionship with people (pleasant, predictable interactions on the cat's own terms) • Opportunities to play (by the cat's self, with the human, and with other cats if part of the same social group) • Reward-based training; no use of aversive methods • Opportunities for enrichment

MENTAL STATE
• A sense of safety • A routine • Prompt help for behavior issues • Choices and agency

This book includes a checklist for a happy cat that will help you to apply some of the ideas from the book to your own life (see page 236). It is only a guideline. If you have any concerns about your cat, see your veterinarian or a suitably qualified cat behaviorist, as appropriate. I hope the list will help you identify things that you are already getting right, so that you know to keep doing them. Most likely you will find some things that you could also do differently, and it's up to you to decide which ones to try. Remember that cats don't like change and you don't have to try lots of things all at once. Pick one thing that you think will work for both you and your cat, and give it a go. Make a note of how your cat responds. For example, if providing a new enrichment item, do they use it? If not, what can you do to encourage them to use it? It could be that making toys a bit easier, using treats, or adding something that smells like the cat will get them to pay attention. Then when you're ready, pick something else off the list and try that one.

If you keep notes about how your cat is doing—whether written notes or little videos of your cat playing, sleeping, and so on—then you'll be able to track whether you think the changes are a good thing from your cat's point of view. It's not possible to make everything perfect for your cat, just as it's not possible for us to live a charmed life, so you may think of some things but decide not to try them, and that's okay too. Ultimately, I hope this book will give you a better understanding of what your cat needs, and by providing it, a better relationship with your cat.

CITIZEN SCIENCE: GET INVOLVED

AS WELL AS being able to make life better for our own cats, the increased interest in cats from scientists means you can help expand our collective knowledge if you would like to. Many researchers need volunteers to fill out questionnaires about you and your cat, and if you live in a city with a university where feline science takes place, scientists might be looking either to test your cat in your home or invite you to the lab to take part in studies there (a more difficult proposition suited only to certain types of cat). You can follow many feline scientists and cat advocacy organizations on social media, where you will find many great stories about what we're learning and what cats want.

Some organizations I'd like to give a special shout-out to are International Cat Care, Cats Protection, the BC SPCA, and the RSPCA, all of which put out lots of great content. The American College of Veterinary Behaviorists, the American Association of Feline Practitioners, and the American Veterinary Society of Animal Behavior also work hard to provide good information on cats. And many of the feline experts from different fields who are quoted or mentioned in this book have active feeds on social media that are well worth following.

Of course, I also hope you will come and find me at my blog, *Companion Animal Psychology*, where I continue to write about what science tells us about cats and dogs, and evidence-based ways to take care of them. You can follow by email or social media as you wish. My blog was inspired by my own pets, Ghost and Bodger the dogs, and Harley and Melina the cats, and it's been a privilege to write about them and share what I've learned.

I've been thinking about what would be Harley's and Melina's perfect days. Harley is a creature of habit, so for him, everything would happen exactly on time. Mealtimes, brushing time, play— even the time at which he gets his night-time treat and we go to bed, because that's when he jumps on the bed for some fuss and then sleep. He loves to be warm, so I think his perfect day would be one in which the heat was on and he could relax on his back on the heat vent, legs akimbo. But he also loves to lie in the sun and watch the hummingbirds, so his perfect day could also be a sunny day when the hummingbirds have arrived. He would lounge in the warm sun by the bedroom window, watching them fly to and from the feeder. He would like to play with one of the wand toys or to follow the laser light up and down the hallway, and he wouldn't mind which. And his day would include several pauses for petting. Plus, he would like to jump on my desk and actually chew on the wires of my headset without being removed.

Melina is more energetic than Harley and she is also not quite as fussy about the precise timings of things (or if she is, she is polite enough not to holler about it). After breakfast, she would retreat to Bodger's old bed for her morning nap. Her perfect day would include some play by herself, like playing football with one of her ball toys or little springs, and also some leaping about after the wand toy. She prefers the wand toy with a pretty realistic butterfly on the end of it. The only wildlife she would see through the window would be birds and squirrels, and she wouldn't hear any dogs bark in the street. In the evening, she would alternate between sitting on my husband's lap and sitting next to me on the settee. She stands on my lap to sniff noses and accept a few pets under the chin, but prefers to sit next to me rather than on my lap. And at night she would cuddle up with Harley and us on the bed.

I think we all want to give our cats as many perfect days as possible. Spending more time with our cats and paying attention to their needs will help them, and us, be happier.

CHECKLIST FOR A HAPPY CAT

THIS CHECKLIST IS designed to help you think about some of the ideas in this book in relation to your own cat. It is not a scientifically validated instrument, nor is it a substitute for a professional opinion. If you have any concerns about your cat, consult a veterinarian (and a behaviorist, if appropriate).

Answer each question "yes" or "no." The more "yes" answers, the better. For the "no" answers, troubleshoot the situation to see if you should make changes (e.g., making food-puzzle toys easier, making something less stressful, etc.).

Cat's Name:
Age:
Breed:

		YES/ NO	CHAPTER
1	My cat's schedule (e.g., mealtimes, playtimes) is pretty much the same every day (i.e., things happen at the same time each day).		1
2	My cat has a safe space in every room.		3
3	My cat has plenty of high-up spaces (such as cat trees, shelves, etc.) throughout the home.		3
4	I make time to play with my cat every day (e.g., with wand toys).		3
5	My cat has toys they play with on their own, and I rotate the toys regularly so my cat doesn't get bored.		3, 7
6	My cat has food-puzzle toys (and when first introduced, they are made nice and easy).		4, 10
7	I use scent as enrichment for my cat (e.g., having toys with catnip/valerian/honeysuckle/silver vine, bringing cat-friendly items in from outside).		7
8	I and other family members try to ensure interactions with the cat are predictable, short, and (if the cat likes it) frequent.		3, 8
9	If very young children want to pet my cat, an adult guides their hand to ensure they are gentle.		8

		YES/ NO	CHAPTER
10	My cat always has a choice of whether or not to be petted.		8
11	I never use spray bottles, shake cans, or other forms of punishment with my cat.		5
12	My cat's litter boxes are in quiet locations away from the cat's other resources.		3, 11
13	I scoop the litter boxes at least once a day and clean them thoroughly at least once a week.		11
14	My cat's litter boxes are nice and big.		11
15	My cat's food and water bowls are separate from other resources.		3
16	My cat has scratching posts that are sturdy, stable, and nice and tall.		3
17	My cat also has a horizontal scratching surface.		3
18	I try to avoid the use of strong scents (such as cleaning products) in places my cat likes to be.		3
19	My cat is brushed regularly and I wipe their eyes and face each day (as needed, depending on the breed).		6
20	I have trained my cat to like their cat carrier, and the carrier is out in the house for them to relax in if they wish.		6

		YES/NO	CHAPTER
21	My cat has annual vet visits (or as often as the vet suggests).		6
22	My cat is vaccinated as per my veterinarian's recommendation.		6
23	My cat has anti-flea and worm treatments as per my veterinarian's recommendation.		6
24	My cat is a healthy weight.		10
25	I look out for signs of stress in my cat and intervene to help if needed.		4, 11
26	If needed, I have made changes to the environment (such as ramps or steps) to accommodate my cat's special needs.		12
27	I have made plans for my cat in case something happens to me.		13
28	My cat is included in my family's emergency planning.		13
	My cat's favorite places to sleep are:		
	My cat's favorite toys are:		
	My cat's best hiding place (where they love to hide and I will not disturb them) is:		

APPENDIX: TRAINING PLANS

TEACH YOUR CAT TO SIT PRETTY

You will need...

Primary reinforcer: Food that your cat likes. Something small (about one-quarter to one-half of the size of your fingernail) and very tasty. Little bits of chicken or tuna are a good idea, but you can use canned wet food or other types of meat, fish, or treats if you wish. Cut them up into the right size in advance, and keep them on or near your person (e.g., in a bait bag or in a bowl near where you will train).

Secondary reinforcer: A clicker. Sometimes, a fearful cat will startle at the sound of the clicker, in which case you could try a quieter clicker (like the i-Click), muffle it in a sock, use the click of a ballpoint pen, or use your voice instead. If you decide to use your voice as your secondary reinforcer, pick a word (e.g., "yes" or "good") and try to say it in the same tone of voice every time.

A **quiet place to train**, and a surface that isn't slippery for the cat. If you don't want to get down to the floor, you could have the cat on a table.

Training tips

- When the cat does the behavior you are looking for, use your secondary reinforcer right away—i.e., click. Then give the cat one of the pieces of chicken or tuna.

- If you accidentally click, you still have to give the reinforcement. The mistake was yours, not the cat's! Try to get your timing better next time.

- Training can be fun for you and the cat, but keep each session short—five minutes is enough.

- To prevent the cat from becoming overweight, keep track of how much food they are getting during training sessions, and reduce the amount they get at mealtimes to compensate.

- Be patient. Sometimes the cat will seem to have "gotten it" and then go backwards in the training plan. This is normal—learning a new behavior takes time, especially if the cat is new to training.

- The training plan has several steps. Make sure the cat is fluent at one step before moving to the next one. If you get stuck and keep going forwards and backwards in the plan, you can use a "split"—a halfway step—as suggested in the plan. This will help you keep up a high rate of reinforcement, which in turn will keep the cat interested.

The training plan

Step 1: Lure the cat into a sit. Do this by putting a piece of food between your finger and thumb, then put the food in front of the cat's nose and lift it up and back slightly. As the cat's nose goes up to follow the food, their rear end will go down to the floor. Keep the food very close to their nose and then once the cat's bottom is on the floor, you can mark the behavior (i.e., click) and give them the food.

Possible split: Get the cat to follow the food just a little bit with their nose (i.e., a partial sit rather than a full sit).

Step 2: Fade the lure. Without any food in your hand, move your hand exactly as in step 1 to move the cat into a sit. The cat's nose should target your hand and follow it up. As soon as the cat's bottom touches the floor, click and give the food.

Possible splits: If the cat is not interested when the food is not in your hand, try these splits to help you fade the lure:

a. Pretend to get a treat in your hand, and use the "pretend treat" to lure the cat. After the click, let them see there was no food in your hand and feed with your other hand.

b. If your cat did not follow the "pretend treat," repeat step 1, but bury the food completely between your finger and thumb so it is less "obvious" to the cat.

c. Repeat step 1, but this time as well as burying the food between finger and thumb, feed with your other hand after you have clicked or said "yes." This way the cat is learning that they do not get the food in your hand, but they do still get food for doing the behavior. Once the cat can do this, go back to step 2.

Step 3: As step 2, but this time, once the cat is in a sit, raise your hand vertically 5 cm (2 inches) above the cat's nose so they have to reach up a bit more. The front paws should naturally come up too. As soon as the cat moves up so their nose is by your hand, click and feed.

Possible split: Raise your hand just 2–3 cm (1 inch) instead so the cat does not have to lift up quite as far.

Step 4: Do the hand signal close to the cat but without your hand being directly by the cat's nose. If your cat finds this hard, do the hand signal a bit closer to the cat but still not right by the nose. The benefit of moving the signal farther away is that you can get the cat to sit pretty without being right in front of them. As soon as the cat sits pretty, click and feed.

Step 5: Once the cat is very good at step 3, add the verbal cue like this: Say "Sit pretty," wait one second, then do the hand signal. Just like before, as soon as the cat does the behavior, click and feed them.

Step 6: Once the cat is fluent at step 5, they will start to sit pretty even before you do the hand signal. When you are really sure the cat will sit on the verbal cue "Sit pretty," you can stop using the hand signal.

Troubleshooting

- You may need to vary the type of food used as a reinforcement so the cat does not get bored. If your cat is not interested in the food, you could try a different type of food, e.g., switch from meat to fish. Every cat is an individual, so you have to find what motivates them.

- If the cat tries to bite or swipe you, you may find it easier to deliver food with a spatula, spoon, syringe, or squeeze tube. This makes it easier for the cat to take the food without touching your fingers. You could also use a target stick instead of your hand as the signal.

- If you are not quite getting the cat into the position you are looking for, pay attention to where the lure or your hand is. Also, if you are too slow to click/reward, the cat may start to reach up with their paws, which you may or may not want. This is a new skill for you and the cat and you can both have fun practicing!

TEACH YOUR CAT TO LIKE THEIR CARRIER

MOST CATS WILL already have a negative association with their carrier, which means you need to make new, positive associations with it as a place to relax and eat treats in. Find a place in your house where the cat carrier can live all the time and be a place that your cat might enjoy.

Before you get started, put a nice towel or fleecy blanket (preferably one that already smells of your cat) in the base of the carrier so that it will be comfortable.

As well, identify something your cat really loves and that you can use, such as pieces of tuna, cat treats, or a short brushing session (only if your cat loves to be brushed). See the section on primary reinforcers in the previous plan for ideas.

It's important to break the training down into very small steps. Work through each of the following steps in order. Proceed at the cat's pace, and don't move on to another step until you are sure your cat is happy with the current step.

If you have multiple cats, they may need to work at different paces (as well as each needing their own carrier, of course). If your cat is not afraid of the carrier, or is a kitten, you will work through the first steps very quickly. But if you see any signs of stress or your cat becomes reluctant to complete that step, go back to the previous step (or a few steps earlier if needed).

The training plan

At each step, reward your cat liberally. If you are feeding treats, feed extra treats while the cat is in the carrier.

You may find that sometimes you advance a step and then have to drop back; this is perfectly normal. Just like for us, learning takes practice!

Once the cat is happy to approach the carrier, you can add a bonus by leaving a nice food reward or a toy in the carrier for them to find. You will see them going to check out the carrier to look for a treat. (Make sure you keep replenishing, as you don't want them to be disappointed!)

STEP NO.	WHAT YOU WANT THE CAT TO DO	WHAT TO DO WITH THE CARRIER	WHERE AND HOW TO REWARD THE CAT
1	Go into the same room as the carrier.	Put nice comfy bedding into the base of the carrier. Put the base of the carrier in a room the cat often uses.	Feed the cat for going into the room; feed at whatever distance from the carrier the cat is comfortable with.
2	Approach the base of the carrier.	As above.	Feed the cat at a distance where they are comfortable, which initially may be far away. Gradually have the cat come closer to be fed.
3	Go onto the base of the carrier.	As above.	Put the cat's rewards onto the base of the carrier. Keep feeding them for as long as they stay.
4	Go into the carrier with the top on.	Put the top on the carrier. Make sure the door stays open.	As above. Start by feeding at the front. Then, gradually put the treats near the back of the carrier to encourage the cat to go fully in.
5	Go into the carrier and stay there while the door is closed for one second and then opened.	When the cat goes in, close the door and then immediately open it again.	Give the cat treats, and if they continue to stay in there after the door is opened, continue to feed.

STEP NO.	WHAT YOU WANT THE CAT TO DO	WHAT TO DO WITH THE CARRIER	WHERE AND HOW TO REWARD THE CAT
6	Go into the carrier and stay there while the door is closed for longer, gradually building up to 30 seconds.	When the cat goes into the carrier, close the door, keep it shut a little longer, then open it. As you gradually build up to 30 seconds (s) with the door closed, keep including some very short closures. For example, try closing for 3s, 5s, 1s, 7s, 10s, 2s, 12s, 1s, 5s, 15s, 3s, 12s, 18s, 2s, 5s, 21s, 3s, 25s, 1s, 9s, 28s, 12s, 30s, 1s.	Continue to give the cat treats all the time they are in the carrier. Don't be stingy!
7	Go into the carrier and stay.	When the cat goes in, close the door, pick up the carrier, then immediately put it down and open the door.	As above.
8	Go into the carrier while it is put in the car and brought back out.	When the cat goes in, close the door securely, take the carrier and put it in the car, then immediately bring it back out and open the door.	As above.
9	Go into the carrier for a very short car ride, which ends back at home.	When the cat goes in the carrier, take it to the car, secure it, and go for a very short drive. This might be only a few meters and then back. You can gradually increase the distance. As splits (if needed), have the carrier in the car while the engine is turned on and then off, then while the indicator is turned on and then off, etc., to gradually build up to a car ride of a few minutes.	As above. You will need someone to help you at this stage! Either someone to drive while you sit in the back of the car and feed the cat, or someone to sit in the back and feed the cat while you drive. If you get someone else to do the feeding, remind them not to be stingy and to keep feeding throughout.

NOTES

Chapter 1: Happy Cats

1   Kristopher Poole, "The contextual cat: Human–animal relations and social meaning in Anglo-Saxon England," *Journal of Archaeological Method and Theory* 22, no. 3 (2015): 857–882.

2   A.F. Haruda et al., "The earliest domestic cat on the Silk Road," *Scientific Reports* 10, no. 1 (2020): 1–12.

3   Jonathan Balcombe, *What a Fish Knows: The Inner Lives of Our Underwater Cousins* (New York: Scientific American/Farrar, Strauss and Giroux, 2016).

4   Annika Stefanie Reinhold et al., "Behavioral and neural correlates of hide-and-seek in rats," *Science* 365, no. 6458 (2019): 1180–1183.

5   Philip Low et al., *The Cambridge Declaration on Consciousness,* Francis Crick Memorial Conference, Cambridge, England, 2012, fcm conference.org/img/CambridgeDeclarationOnConsciousness.pdf.

6   John Bradshaw, "Normal feline behaviour: ... and why problem behaviours develop," *Journal of Feline Medicine and Surgery* 20, no. 5 (2018): 411–421.

7   Kristyn R. Vitale Shreve and Monique A.R. Udell, "What's inside your cat's head? A review of cat (*Felis silvestris catus*) cognition research past, present and future," *Animal Cognition* 18, no. 6 (2015): 1195–1206.

8   David J. Mellor, "Updating animal welfare thinking: Moving beyond the 'Five Freedoms' towards 'a life worth living,'" *Animals* 6, no. 3 (2016): 21; David J. Mellor, "Moving beyond the 'Five Freedoms' by updating the 'Five Provisions' and introducing aligned 'animal welfare aims,'" *Animals* 6, no. 10 (2016): 59.

9   Mellor, "Moving beyond the 'Five Freedoms'"; David J. Mellor et al., "The 2020 Five Domains Model: Including human–animal interactions in assessments of animal welfare," *Animals* 10, no. 10 (2020): 1870.

10 Jean-Loup Rault et al., "The power of a positive human–animal relationship for animal welfare," *Frontiers in Veterinary Science* 7 (2020).

11 Mellor et al., "The 2020 Five Domains Model."

12 People's Dispensary for Sick Animals, *PDSA Animal Wellbeing (PAW) Report: The Essential Insight into the Wellbeing of UK Pets*, 2020, pdsa.org .uk/media/10509/20039_pdsa-paw-report-2020_7-10_press_3 _online-5.pdf.

13 Fiona Rioja-Lang et al., "Determining priority welfare issues for cats in the United Kingdom using expert consensus," *Veterinary Record Open* 6, no. 1 (2019).

14 Emma K. Grigg and Lori R. Kogan, "Owners' attitudes, knowledge, and care practices: Exploring the implications for domestic cat behavior and welfare in the home," *Animals* 9, no. 11 (2019): 978.

15 Tiffani J. Howell, Kate Mornement, and Pauleen C. Bennett, "Pet cat management practices among a representative sample of owners in Victoria, Australia," *Journal of Veterinary Behavior* 11 (2016): 42–49.

16 Emma K. Grigg et al., "Cat owners' perceptions of psychoactive medications, supplements and pheromones for the treatment of feline behavior problems," *Journal of Feline Medicine and Surgery* 21, no. 10 (2019): 902–909.

17 Jaak Panksepp, "Affective consciousness: Core emotional feelings in animals and humans," *Consciousness and Cognition* 14, no. 1 (2005): 30–80.

18 International Cat Care, "Top tip: Understanding cat blinks," 2020, icatcare.org/top-tip-understanding-cat-blinks.

19 Nadine Gourkow and Clive J.C. Phillips, "Effect of cognitive enrichment on behavior, mucosal immunity and upper respiratory disease of shelter

cats rated as frustrated on arrival," *Preventive Veterinary Medicine* 131 (2016): 103–110.

20 Valerie Bennett, Nadine Gourkow, and Daniel S. Mills, "Facial correlates of emotional behaviour in the domestic cat (*Felis catus*)," *Behavioural Processes* 141 (2017): 342–350.

21 Marina C. Evangelista et al., "Facial expressions of pain in cats: The development and validation of a Feline Grimace Scale," *Scientific Reports* 9, no. 1 (2019): 1–11.

22 Lauren C. Dawson et al., "Humans can identify cats' affective states from subtle facial expressions," *Animal Welfare* 28, no. 4 (2019): 519–531.

Chapter 2: Getting a Kitten or Cat

1 RSPCA, "Kittens for sale," n.d., rspca.org.uk/adviceandwelfare/pets /cats/kittens.

2 S.L. Crowell-Davis, T.M. Curtis, and R.J. Knowles, "Social organization in the cat: A modern understanding," *Journal of Feline Medicine and Surgery* 6 (2004): 19–28.

3 Lauren R. Finka et al., "Owner personality and the wellbeing of their cats share parallels with the parent-child relationship," *PLOS ONE* 14, no. 2 (2019): e0211862.

4 Mark J. Farnworth et al., "Flat feline faces: Is brachycephaly associated with respiratory abnormalities in the domestic cat (*Felis catus*)?," *PLOS ONE* 11, no. 8 (2016): e0161777.

5 Kerstin L. Anagrius et al., "Facial conformation characteristics in Persian and Exotic Shorthair cats," *Journal of Feline Medicine and Surgery* (2021): 1098612X21997631.

6 Lauren R. Finka et al., "The application of geometric morphometrics to explore potential impacts of anthropocentric selection on animals' ability to communicate via the face: The domestic cat as a case study," *Frontiers in Veterinary Science* 7 (2020): 1070.

7 International Cat Care, "Bengal," 2018, icatcare.org/advice/bengal.

8 Mark J. Farnworth et al., "In the eye of the beholder: Owner preferences for variations in cats' appearances with specific focus on skull morphology," *Animals* 8, no. 2 (2018): 30.

9   Jacqueline Wilhelmy et al., "Behavioral associations with breed, coat type, and eye color in single-breed cats," *Journal of Veterinary Behavior* 13 (2016): 80–87.

10 Milla Salonen et al., "Breed differences of heritable behaviour traits in cats," *Scientific Reports* 9, no. 1 (2019): 1–10.

11 Roberta R. Collard, "Fear of strangers and play behavior in kittens with varied social experience," *Child Development* (1967): 877–891; John W.S. Bradshaw, Rachel A. Casey, and Sarah L. Brown, *The Behaviour of the Domestic Cat* (Boston, MA: CABI, 2012).

12 Michael W. Fox (1970), as cited in Dennis C. Turner, "A review of over three decades of research on cat-human and human-cat interactions and relationships," *Behavioural Processes* 141 (2017): 297–304.

13 Thomas McNamee, *The Inner Life of Cats: The Science and Secrets of Our Mysterious Feline Companions* (New York: Hachette, 2018).

14 Milla K. Ahola, Katariina Vapalahti, and Hannes Lohi, "Early weaning increases aggression and stereotypic behaviour in cats," *Scientific Reports* 7, no. 1 (2017): 1–9.

15 ASPCA Pro, "Telling a kitten's age in four steps," n.d., aspcapro.org /resource/telling-kittens-age-four-steps.

Chapter 3: How to Set Up Your Home for a Cat

1   Fiona Rioja-Lang et al., "Determining priority welfare issues for cats in the United Kingdom using expert consensus," *Veterinary Record Open* 6, no. 1 (2019).

2   S.L. Ellis et al., "AAFP and ISFM feline environmental needs guidelines," *Journal of Feline Medicine and Surgery* 15, no. 3 (2013): 219–230.

3   J.J. Ellis et al., "Environmental enrichment choices of shelter cats," *Behavioural Processes* 141, no. 3 (2017): 291–296.

4   Sarah L. Hall, John W.S. Bradshaw, and Ian H. Robinson, "Object play in adult domestic cats: The roles of habituation and disinhibition," *Applied Animal Behaviour Science* 79, no. 3 (2002): 263–271.

5   B. Strickler and E. Shull, "An owner survey of toys, activities, and behavior problems in indoor cats," *Journal of Veterinary Behavior: Clinical Applications and Research* 9, no. 5 (2014): 207–214.

6   K.R.V. Shreve and M.A. Udell, "Stress, security, and scent: The influence of chemical signals on the social lives of domestic cats and implications for applied settings," *Applied Animal Behaviour Science* 187 (2017): 69–76.

7   Mei S. Yamaguchi et al., "Bacteria isolated from Bengal cat (*Felis catus × Prionailurus bengalensis*) anal sac secretions produce volatile compounds potentially associated with animal signaling," *PLOS ONE* 14, no. 9 (2019): e0216846.

8   Miyabi Nakabayashi, Ryohei Yamaoka, and Yoshihiro Nakashima, "Do faecal odours enable domestic cats (*Felis catus*) to distinguish familiarity of the donors?," *Journal of Ethology* 30, no. 2 (2012): 325–329.

9   Manuel Mengoli et al., "Scratching behaviour and its features: A questionnaire-based study in an Italian sample of domestic cats," *Journal of Feline Medicine and Surgery* 15, no. 10 (2013): 886–892.

10 Colleen Wilson et al., "Owner observations regarding cat scratching behavior: An internet-based survey," *Journal of Feline Medicine and Surgery* 18, no. 10 (2016): 791–797.

11 Lingna Zhang, Rebekkah Plummer, and John McGlone, "Preference of kittens for scratchers," *Journal of Feline Medicine and Surgery* 21, no. 8 (2019): 691–699.

Chapter 4: Key Aspects of Caring for a Cat

1   Catherine M. Hall et al., "Factors determining the home ranges of pet cats: A meta-analysis," *Biological Conservation* 203 (2016): 313–320.

2   Peter Sandøe et al., "The burden of domestication: A representative study of welfare in privately owned cats in Denmark," *Animal Welfare* 26 (2017): 1–10.

3   Daiana de Souza Machado et al., "Beloved whiskers: Management type, care practices and connections to welfare in domestic cats," *Animals* 10, no. 12 (2020): 2308.

4   Sarah M.L. Tan, Anastasia C. Stellato, and Lee Niel, "Uncontrolled outdoor access for cats: An assessment of risks and benefits," *Animals* 10, no. 2 (2020): 258.

5   RSPCA, "What are the signs of antifreeze poisoning in cats?," n.d., rspca.org.uk/adviceandwelfare/pets/cats/health/poisoning/antifreeze.

6   I. Rochlitz, " The effects of road traffic accidents on domestic cats and their owners," *Animal Welfare* 13, no. 1 (2004): 51–56.

7   Stanley D. Gehrt et al., "Population ecology of free-roaming cats and interference competition by coyotes in urban parks," *PLOS ONE* 8, no. 9 (2013): e75718.

8   Rachel N. Larson et al., "Effects of urbanization on resource use and individual specialization in coyotes (*Canis latrans*) in southern California," *PLOS ONE* 15, no. 2 (2020): e0228881.

9   S.A. Poessel, E.C. Mock, and S.W. Breck, "Coyote (*Canis latrans*) diet in an urban environment: Variation relative to pet conflicts, housing density, and season," *Canadian Journal of Zoology* 95, no. 4 (2017): 287–297; E. MacDonald, T. Milfont, and M. Gavin, "What drives cat-owner behaviour? First steps towards limiting domestic-cat impacts on native wildlife," *Wildlife Research* 42, no. 3 (2015): 257–265.

10 Royal Society for the Protection of Birds, "How many birds do cats kill," n.d., rspb.org.uk/birds-and-wildlife/advice/gardening-for-wildlife/animal-deterrents/cats-and-garden-birds/are-cats-causing-bird-declines.

11 Roland W. Kays and Amielle A. DeWan, "Ecological impact of inside/outside house cats around a suburban nature preserve," *Animal Conservation* 7, no. 3 (2004): 273–283.

12 Michael Calver et al., "Reducing the rate of predation on wildlife by pet cats: The efficacy and practicability of collar-mounted pounce protectors," *Biological Conservation* 137, no. 3 (2007): 341–348.

13 Catherine M. Hall et al., "Assessing the effectiveness of the Birdsbesafe® anti-predation collar cover in reducing predation on wildlife by pet cats in Western Australia," *Applied Animal Behaviour Science* 173 (2015): 40–51.

14 Martina Cecchetti et al., "Provision of high meat content food and object play reduce predation of wild animals by domestic cats *Felis catus*," *Current Biology* 31, no. 5 (2021): 1107–1111.e5.

15 Tiffani J. Howell, Kate Mornement, and Pauleen C. Bennett, "Pet cat management practices among a representative sample of owners in Victoria, Australia," *Journal of Veterinary Behavior* 11 (2016): 42–49.

16 Scott S. Campbell and Irene Tobler, "Animal sleep: A review of sleep duration across phylogeny," *Neuroscience & Biobehavioral Reviews* 8, no. 3 (1984): 269-300.

17 John W.S. Bradshaw, Rachel A. Casey, and Sarah L. Brown, *The Behaviour of the Domestic Cat* (Boston, MA: CABI, 2012).

18 Daniel E. Slotnik, "Michel Jouvet, who unlocked REM's sleep secrets, dies at 91," *New York Times*, 2017, nytimes.com/2017/10/11/obituaries/michel-jouvet-who-unlocked-rem-sleeps-secrets-dies-at-91.html; Barbara E. Jones, "The mysteries of sleep and waking unveiled by Michel Jouvet," *Sleep Medicine* 49 (2018): 14-19.

19 Daoyun Ji and Matthew A. Wilson, "Coordinated memory replay in the visual cortex and hippocampus during sleep," *Nature Neuroscience* 10, no. 1 (2007): 100-107.

20 Christy L. Hoffman, Kaylee Stutz, and Terrie Vasilopoulos, "An examination of adult women's sleep quality and sleep routines in relation to pet ownership and bedsharing," *Anthrozoös* 31, no. 6 (2018): 711-725.

21 Giuseppe Piccione et al., "Daily rhythm of total activity pattern in domestic cats (*Felis silvestris catus*) maintained in two different housing conditions," *Journal of Veterinary Behavior* 8, no. 4 (2013): 189-194.

Chapter 5: How to Train a Cat

1 Pamela J. Reid, *Excel-erated Learning: Explaining in Plain English How Dogs Learn and How Best to Teach Them* (Berkeley, CA: James and Kenneth Publishers, 2011).

2 N. Porters et al., "Development of behavior in adopted shelter kittens after gonadectomy performed at an early age or at a traditional age," *Journal of Veterinary Behavior: Clinical Applications and Research* 9, no. 5 (2014), 196-206.

3 Kristina A. O'Hanley, David L. Pearl, and Lee Niel, "Risk factors for aggression in adult cats that were fostered through a shelter program as kittens," *Applied Animal Behaviour Science* 236 (2021): 105251.

4 John W.S. Bradshaw, Rachel A. Casey, and Sarah L. Brown, *The Behaviour of the Domestic Cat* (Boston, MA: CABI, 2012).

5 Claudia Fugazza et al., "Did we find a copycat? Do as I do in a domestic cat (*Felis catus*)," *Animal Cognition* 24 (2020): 121-131.

6   L. Pratsch et al., "Carrier training cats reduces stress on transport to a veterinary practice," *Applied Animal Behaviour Science* 206 (2018): 64–74.

7   J. Lockhart, K. Wilson, and C. Lanman, "The effects of operant training on blood collection for domestic cats," *Applied Animal Behaviour Science* 143, no. 2–4 (2013): 128–134.

8   L. Kogan, C. Kolus, and R. Schoenfeld-Tacher, "Assessment of clicker training for shelter cats," *Animals* 7, no. 10 (2017): 73.

9   N. Gourkow and C. Phillips, "Effect of cognitive enrichment on behavior, mucosal immunity and upper respiratory disease of shelter cats rated as frustrated on arrival," *Preventive Veterinary Medicine* 131 (2016): 103–110.

Chapter 6: The Vet and Grooming

1   John O. Volk et al., "Executive summary of the Bayer veterinary care usage study," *Journal of the American Veterinary Medical Association* 238, no. 10 (2011): 1275–1282.

2   John O. Volk et al., "Executive summary of phase 2 of the Bayer veterinary care usage study," *Journal of the American Veterinary Medical Association* 239, no. 10 (2011): 1311–1316.

3   Zoe Belshaw et al., "Owners and veterinary surgeons in the United Kingdom disagree about what should happen during a small animal vaccination consultation," *Veterinary Sciences* 5, no. 1 (2018): 7.

4   Carly M. Moody et al., "Can you handle it? Validating negative responses to restraint in cats," *Applied Animal Behaviour Science* 204 (2018): 94–100.

5   Carly M. Moody et al., "Getting a grip: Cats respond negatively to scruffing and clips," *Veterinary Record* 186, no. 12 (2020): 385.

6   C.M. Moody, C.E. Dewey, and L. Niel, "Cross-sectional survey of cat handling practices in veterinary clinics throughout Canada and the United States," *Journal of the American Veterinary Medical Association* 256, no. 9 (2020): 1020–1033.

7   Chiara Mariti et al., "Guardians' perceptions of cats' welfare and behavior regarding visiting veterinary clinics," *Journal of Applied Animal Welfare Science* 19, no. 4 (2016): 375–384.

8   Stefanie Riemer et al., "A review on mitigating fear and aggression in dogs and cats in a veterinary setting," *Animals* 11, no. 1 (2021): 158.

9   American Association of Feline Practitioners, "Cat friendly homes," n.d., catfriendly.com.

10 Amy E.S. Stone et al., "2020 AAHA/AAFP feline vaccination guidelines," *Journal of Feline Medicine and Surgery* 22, no. 9 (2020): 813–830.

11 Jan Bellows et al., "2019 AAHA dental care guidelines for dogs and cats," *Journal of the American Animal Hospital Association* 55, no. 2 (2019): 49–69.

12 Marianne Diez et al., "Health screening to identify opportunities to improve preventive medicine in cats and dogs," *Journal of Small Animal Practice* 56, no. 7 (2015): 463–469.

13 D.G. O'Neill et al., "Prevalence of disorders recorded in cats attending primary-care veterinary practices in England," *Veterinary Journal* 202, no. 2 (2014): 286–291.

14 N.C. Finch, H.M. Syme, and J. Elliott, "Risk factors for development of chronic kidney disease in cats," *Journal of Veterinary Internal Medicine* 30, no. 2 (2016): 602–610.

15 S.R. Urfer et al., "Risk factors associated with lifespan in pet dogs evaluated in primary care veterinary hospitals," *Journal of the American Animal Hospital Association* 55, no. 3 (2019): 130–137.

Chapter 7: Enrichment for Cats

1   Sarah Ellis, "Environmental enrichment: Practical strategies for improving feline welfare," *Journal of Feline Medicine and Surgery* 11 (2009): 901–912.

2   M.R. Shyan-Norwalt, "Caregiver perceptions of what indoor cats do 'for fun,'" *Journal of Applied Animal Welfare Science* 8, no. 3 (2005): 199–209.

3   Sarah L.H. Ellis and Deborah L. Wells, "The influence of visual stimulation on the behaviour of cats housed in a rescue shelter," *Applied Animal Behaviour Science* 113, no. 1–3 (2008): 166–174.

4   Sarah L.H. Ellis and Deborah L. Wells, "The influence of olfactory stimulation on the behaviour of cats housed in a rescue shelter," *Applied Animal Behaviour Science* 123, no. 1–2 (2010): 56–62.

5   Neil B. Todd, "Inheritance of the catnip response in domestic cats," *Journal of Heredity* 53, no. 2 (1962): 54–56.

6   Benjamin R. Lichman et al., "The evolutionary origins of the cat attractant nepetalactone in catnip," *Science Advances* 6, no. 20 (2020): eaba0721.

7   Reiko Uenoyama et al., "The characteristic response of domestic cats to plant iridoids allows them to gain chemical defence against mosquitoes," *Science Advances* 7, no. 4 (2021): eabd9135.

8   S. Bol et al., "Responsiveness of cats (Felidae) to silver vine (*Actinidia polygama*), Tatarian honeysuckle (*Lonicera tatarica*), valerian (*Valeriana officinalis*) and catnip (*Nepeta cataria*)," BMC *Veterinary Research* 13, no. 1 (2017): 70.

9   John W.S. Bradshaw, Rachel A. Casey, and Sarah L. Brown, *The Behaviour of the Domestic Cat* (Boston, MA: CABI, 2012).

10 Charles T. Snowdon, David Teie, and Megan Savage, "Cats prefer species-appropriate music," *Applied Animal Behaviour Science* 166 (2015): 106–111.

11 Amanda Hampton et al., "Effects of music on behavior and physiological stress response of domestic cats in a veterinary clinic," *Journal of Feline Medicine and Surgery* 22, no. 2 (2020): 122–128.

12 Emily G. Patterson-Kane and Mark J. Farnworth, "Noise exposure, music, and animals in the laboratory: A commentary based on Laboratory Animal Refinement and Enrichment Forum (LAREF) discussions," *Journal of Applied Animal Welfare Science* 9, no. 4 (2006): 327–332.

Chapter 8: Cats and Their People

1   Kristyn R. Vitale, Alexandra C. Behnke, and Monique A.R. Udell, "Attachment bonds between domestic cats and humans," *Current Biology* 29, no. 18 (2019): R864–R865.

2   Alice Potter and Daniel Simon Mills, "Domestic cats (*Felis silvestris catus*) do not show signs of secure attachment to their owners," PLOS ONE 10, no. 9 (2015): e0135109.

3   A. Saito et al., "Domestic cats (*Felis catus*) discriminate their names from other words," *Scientific Reports* 9 (2019): 5394.

4   Saho Takagi et al., "Cats match voice and face: Cross-modal represen-
    tation of humans in cats (*Felis catus*)," *Animal Cognition* 22, no. 5 (2019):
    901–906.

5   Dennis C. Turner, "A review of over three decades of research on
    cat-human and human-cat interactions and relationships," *Behavioural
    Processes* 141 (2017): 297–304.

6   Claudia Mertens and Dennis C. Turner, "Experimental analysis of
    human-cat interactions during first encounters," *Anthrozoös* 2, no. 2
    (1988): 83–97.

7   Kristyn R. Vitale and Monique A.R. Udell, "The quality of being sociable:
    The influence of human attentional state, population, and human famil-
    iarity on domestic cat sociability," *Behavioural Processes* 158 (2019):
    11–17.

8   Kristyn R. Vitale Shreve, Lindsay R. Mehrkam, and Monique A.R. Udell,
    "Social interaction, food, scent or toys? A formal assessment of domestic
    pet and shelter cat (*Felis silvestris catus*) preferences," *Behavioural Pro-
    cesses* 141 (2017): 322–328.

9   Matilda Eriksson, Linda J. Keeling, and Therese Rehn, "Cats and own-
    ers interact more with each other after a longer duration of separation,"
    PLOS ONE 12, no. 10 (2017): e0185599.

10 Moriah Galvan and Jennifer Vonk, "Man's other best friend: Domestic
    cats (*F. silvestris catus*) and their discrimination of human emotion cues,"
    Animal Cognition 19, no. 1 (2016): 193–205.

11 Isabella Merola et al., "Social referencing and cat-human communica-
    tion," *Animal Cognition* 18, no. 3 (2015): 639–648.

12 Ádám Miklósi et al., "A comparative study of the use of visual communi-
    cative signals in interactions between dogs (*Canis familiaris*) and humans
    and cats (*Felis catus*) and humans," *Journal of Comparative Psychology* 119,
    no. 2 (2005): 179.

13 Sarah L.H. Ellis, Victoria Swindell, and Oliver H.P. Burman, "Human
    classification of context-related vocalizations emitted by familiar and
    unfamiliar domestic cats: An exploratory study," *Anthrozoös* 28, no. 4
    (2015): 625–634.

14 Karen McComb et al., "The cry embedded within the purr," *Current Biol-
    ogy* 19, no. 13 (2009): R507–R508.

15 Sarah Lesley Helen Ellis et al., "The influence of body region, handler familiarity and order of region handled on the domestic cat's response to being stroked," *Applied Animal Behaviour Science* 173 (2015): 60-67.

16 Lynette A. Hart et al., "Compatibility of cats with children in the family," *Frontiers in Veterinary Science* 5 (2018): 278.

17 John W.S. Bradshaw, "Sociality in cats: A comparative review," *Journal of Veterinary Behavior* 11 (2016): 113-124.

18 Tasmin Humphrey et al., "The role of cat eye narrowing movements in cat-human communication," *Scientific Reports* 10, no. 1 (2020): 1-8.

Chapter 9: The Social Cat

1   S.L. Crowell-Davis, T.M. Curtis, and R.J. Knowles, "Social organization in the cat: A modern understanding," *Journal of Feline Medicine and Surgery* 6, no. 1 (2004): 19-28.

2   John W.S. Bradshaw, "Sociality in cats: A comparative review," *Journal of Veterinary Behavior* 11 (2016): 113-124.

3   Noema Gajdoš Kmecová et al., "Potential risk factors for aggression and playfulness in cats: Examination of a pooling fallacy using Fe-BARQ as an example," *Frontiers in Veterinary Science* 7 (2021): 545326.

4   Rachel Foreman-Worsley and Mark J. Farnworth, "A systematic review of social and environmental factors and their implications for indoor cat welfare," *Applied Animal Behaviour Science* 220 (2019): 104841.

5   Theresa L. DePorter et al., "Evaluation of the efficacy of an appeasing pheromone diffuser product vs placebo for management of feline aggression in multi-cat households: A pilot study," *Journal of Feline Medicine and Surgery* 21, no. 4 (2019): 293-305.

6   As described in an online lecture in 2017 by Dr. Charlotte Cameron-Beaumont, in which she gave information from her doctoral thesis: C.L. Cameron-Beaumont, "Visual and tactile communication in the domestic cat (*Felis silvestris catus*) and undomesticated small felids," PhD thesis, University of Southampton, UK, 1997.

7   S. Cafazzo and E. Natoli, "The social function of tail up in the domestic cat (*Felis silvestris catus*)," *Behavioural Processes* 80, no. 1 (2009): 60-66.

8   John Bradshaw, "Normal feline behaviour: . . . and why problem behaviours develop," *Journal of Feline Medicine and Surgery* 20, no. 5 (2018): 411–421.

9   Mikel Delgado and Julie Hecht, "A review of the development and functions of cat play, with future research considerations," *Applied Animal Behaviour Science* 214 (2019): 1–17; John W.S. Bradshaw, Rachel A. Casey, and Sarah L. Brown, *The Behaviour of the Domestic Cat* (Boston, MA: CABI, 2012).

10 N. Feuerstein and J. Turkel, "Interrelationships of dogs (*Canis familiaris*) and cats (*Felis catus L.*) living under the same roof," *Applied Animal Behaviour Science* 113 (2007): 150–165.

11 J.E. Thomson, S.S. Hall, and D.S. Mills, "Evaluation of the relationship between cats and dogs living in the same home," *Journal of Veterinary Behavior* 27 (2018): 35–40.

12 Miriam Rebecca Prior and Daniel Simon Mills, "Cats vs. dogs: The efficacy of Feliway Friends™ and Adaptil™ products in multispecies homes," *Frontiers in Veterinary Science* 7 (2020): 399.

Chapter 10: Feeding Your Cat

1   Mikel Delgado and Leticia M.S. Dantas, "Feeding cats for optimal mental and behavioral well-being," *Veterinary Clinics: Small Animal Practice* 50, no. 5 (2020): 939–953.

2   Tammy Sadek et al., "Feline feeding programs: Addressing behavioural needs to improve feline health and wellbeing," *Journal of Feline Medicine and Surgery* 20, no. 11 (2018): 1049–1055.

3   A. Alho, J. Pontes, and C. Pomba, "Guardians' knowledge and husbandry practices of feline environmental enrichment," *Journal of Applied Animal Welfare Science* 19, no. 2 (2016): 115–125.

4   L. Dantas et al., "Food puzzles for cats: Feeding for physical and emotional wellbeing," *Journal of Feline Medicine and Surgery* 18, no. 9 (2016).

5   Deborah E. Linder, "Cats are not small dogs: Unique nutritional needs of cats," *Petfoodology* (blog), Cummings Veterinary Medical Center, 2018, vetnutrition.tufts.edu/2018/12/cats-are-not-small-dogs-unique-nutritional-needs-of-cats.

6   J.L. Stella and C.A.T. Buffington, "Individual and environmental effects on cat welfare," ch. 13 in Dennis C. Turner and Patrick Bateson (eds.), *The Domestic Cat: The Biology of Its Behaviour*, 3rd ed. (Cambridge, UK: Cambridge University Press, 2015).

7   Martina Cecchetti et al., "Provision of high meat content food and object play reduce predation of wild animals by domestic cats *Felis catus*," *Current Biology* 31, no. 5 (2021): 1107–1111.e5.

8   Stacie C. Summers et al., "Evaluation of nutrient content and caloric density in commercially available foods formulated for senior cats," *Journal of Veterinary Internal Medicine* 34, no. 5 (2020): 2029–2035.

9   ASPCA Pro, "People foods pets should never eat," n.d., aspcapro.org /resource/people-foods-pets-should-never-eat.

10 Kathryn Michel and Margie Scherk, "From problem to success: Feline weight loss programs that work," *Journal of Feline Medicine and Surgery* 14, no. 5 (2012): 327–336.

11 World Small Animal Veterinary Association, *Body Condition Score*, 2020, wsava.org/wp-content/uploads/2020/02/Body-Condition-Score-Cat .pdf.

12 J. K. Murray et al., "Cohort profile: The 'Bristol Cats Study' (BCS)—A birth cohort of kittens owned by UK households," *International Journal of Epidemiology* 46, no. 6 (2017): 1749–1750e.

13 Elizabeth M. Lund et al., "Prevalence and risk factors for obesity in adult cats from private US veterinary practices," *International Journal of Applied Research in Veterinary Medicine* 3, no. 2 (2005): 88–96.

14 Laurence Colliard et al., "Prevalence and risk factors of obesity in an urban population of healthy cats," *Journal of Feline Medicine and Surgery* 11, no. 2 (2009): 135–140.

15 John Flanagan et al., "An international multi-centre cohort study of weight loss in overweight cats: Differences in outcome in different geographical locations," *PLOS ONE* 13, no. 7 (2018): e0200414.

16 Emily D. Levine et al., "Owner's perception of changes in behaviors associated with dieting in fat cats," *Journal of Veterinary Behavior* 11 (2016): 37–41.

17 Séverine Ligout et al., "Cats reorganise their feeding behaviours when moving from ad libitum to restricted feeding," *Journal of Feline Medicine and Surgery* 22, no. 10 (2020): 953–958.

18 E. Kienzle and R. Bergler, "Human-animal relationship of owners of normal and overweight cats," *Journal of Nutrition* 136, no. 7 (2006): 1947S–1950S.

19 J.B. Coe et al., "Dog owner's accuracy measuring different volumes of dry dog food using three different measuring devices," *Veterinary Record* 185, no. 19 (2019): 599.

20 D. Brooks et al., "2014 AAHA weight management guidelines for dogs and cats," *Journal of the American Animal Hospital Association* 50, no. 1 (2014): 1–11.

Chapter 11: Behavior Problems in Cats

1 S. Scott et al., "Follow-up surveys of people who have adopted dogs and cats from an Australian shelter," *Applied Animal Behaviour Science* 201 (2018).

2 John Bradshaw, "Normal feline behaviour: . . . and why problem behaviours develop," *Journal of Feline Medicine and Surgery* 20, no. 5 (2018): 411–421.

3 Daniel S. Mills et al., "Pain and problem behavior in cats and dogs," *Animals* 10, no. 2 (2020): 318.

4 Emma K. Grigg and Lori R. Kogan, "Owners' attitudes, knowledge, and care practices: Exploring the implications for domestic cat behavior and welfare in the home," *Animals* 9, no. 11 (2019): 978.

5 Sophie Liu et al., "A six-year retrospective study of outcomes of surrendered cats (*Felis catus*) with periuria in a no-kill shelter," *Journal of Veterinary Behavior* 42 (2021): 75–80.

6 Virginie Villeneuve-Beugnet and Frederic Beugnet, "Field assessment of cats' litter box substrate preferences," *Journal of Veterinary Behavior* 25 (2018): 65–70.

7 Norma C. Guy, Marti Hopson, and Raphaël Vanderstichel, "Litterbox size preference in domestic cats (*Felis catus*)," *Journal of Veterinary Behavior* 9, no. 2 (2014): 78–82.

8 Emma K. Grigg, Lindsay Pick, and Belle Nibblett, "Litter box preference in domestic cats: Covered versus uncovered," *Journal of Feline Medicine and Surgery* 15, no. 4 (2013): 280–284.

9   Virginie Villeneuve-Beugnet and Frederic Beugnet, "Field assessment in single-housed cats of litter box type (covered/uncovered) preferences for defecation," *Journal of Veterinary Behavior* 36 (2020): 65–69.

10 J.J. Ellis, R.T.S. McGowan, and F. Martin, "Does previous use affect litter box appeal in multi-cat households?," *Behavioural Processes* 141 (2017): 284–290.

11 Wailani Sung and Sharon L. Crowell-Davis, "Elimination behavior patterns of domestic cats (*Felis catus*) with and without elimination behavior problems," *American Journal of Veterinary Research* 67, no. 9 (2006): 1500–1504.

12 Ragen T.S. McGowan et al., "The ins and outs of the litter box: A detailed ethogram of cat elimination behavior in two contrasting environments," *Applied Animal Behaviour Science* 194 (2017): 67–78.

13 Ana Maria Barcelos et al., "Common risk factors for urinary house soiling (periuria) in cats and its differentiation: The sensitivity and specificity of common diagnostic signs," *Frontiers in Veterinary Science* 5 (2018): 108.

14 Daniela Ramos et al., "A closer look at the health of cats showing urinary house-soiling (periuria): A case-control study," *Journal of Feline Medicine and Surgery* 21, no. 8 (2019): 772–779.

15 Hazel Carney et al., "AAFP and ISFM guidelines for diagnosing and solving house-soiling behavior in cats," *Journal of Feline Medicine and Surgery* 16, no. 7 (2014): 579–598.

16 Nicole K. Martell-Moran, Mauricio Solano, and Hugh G.G. Townsend, "Pain and adverse behavior in declawed cats," *Journal of Feline Medicine and Surgery* 20, no. 4 (2018): 280–288.

17 The Paw Project, n.d., pawproject.org.

18 Ilana R. Reisner et al., "Friendliness to humans and defensive aggression in cats: The influence of handling and paternity," *Physiology & Behavior* 55, no. 6 (1994): 1119–1124.

19 Sandra McCune, "The impact of paternity and early socialisation on the development of cats' behaviour to people and novel objects," *Applied Animal Behaviour Science* 45, no. 1–2 (1995): 109–124.

20 I.C.G. Weaver et al., "Epigenetic programming by maternal behavior," *Nature Neuroscience* 7 (2004): 847–854.

21 Patricia Vetula Gallo, Jack Werboff, and Kirvin Knox, "Development of home orientation in offspring of protein-restricted cats," *Developmental Psychobiology: The Journal of the International Society for Developmental Psychobiology* 17, no. 5 (1984): 437–449.

22 Kristina O'Hanley, David L. Pearl, and Lee Niel, "Risk factors for aggression in adult cats that were fostered through a shelter program as kittens," *Applied Animal Behaviour Science* 236 (2021): 105251.

23 Jorge Palacio et al., "Incidence of and risk factors for cat bites: A first step in prevention and treatment of feline aggression," *Journal of Feline Medicine and Surgery* 9, no. 3 (2007): 188–195.

24 Daniela Ramos and Daniel Simon Mills, "Human directed aggression in Brazilian domestic cats: Owner reported prevalence, contexts and risk factors," *Journal of Feline Medicine and Surgery* 11, no. 10 (2009): 835–841.

25 A.R. Dale et al., "A survey of owners' perceptions of fear of fireworks in a sample of dogs and cats in New Zealand," *New Zealand Veterinary Journal* 58, no. 6 (2010): 286–291.

26 Stephanie Schwartz, "Separation anxiety syndrome in cats: 136 cases (1991–2000)," *Journal of the American Veterinary Medical Association* 220, no. 7 (2002): 1028–1033.

27 Daiana de Souza Machado et al., "Identification of separation-related problems in domestic cats: A questionnaire survey," *PLOS ONE* 15, no. 4 (2020): e0230999.

28 Karen L. Overall, *Manual of Clinical Behavioral Medicine for Dogs and Cats* (Maryland Heights, MO: Mosby, Elsevier, 2013).

29 Emma K. Grigg et al., "Cat owners' perceptions of psychoactive medications, supplements and pheromones for the treatment of feline behavior problems," *Journal of Feline Medicine and Surgery* 21, no. 10 (2019): 902–909.

30 Marta Amat, Tomàs Camps, and Xavier Manteca, "Stress in owned cats: Behavioural changes and welfare implications," *Journal of Feline Medicine and Surgery* 18, no. 8 (2016): 577–586; Debra F. Horwitz and Ilona Rodan, "Behavioral awareness in the feline consultation: Understanding physical and emotional health," *Journal of Feline Medicine and Surgery* 20, no. 5 (2018): 423–436.

Chapter 12: Senior Cats and Cats with Special Needs

1   Michael Ray et al., "2021 AAFP feline senior care guidelines," *Journal of Feline Medicine and Surgery* 23 (2021): 613–638; Jessica Quimby et al., "2021 AAHA/AAFP feline life stage guidelines," *Journal of the American Animal Hospital Association* 57, no. 2 (2021): 51–72.

2   Ray et al., "2021 AAFP feline senior care guidelines"; Quimby et al., "2021 AAHA/AAFP feline life stage guidelines"; Jan Bellows et al., "Aging in cats: Common physical and functional changes," *Journal of Feline Medicine and Surgery* 18, no. 7 (2016): 533–550.

3   Lorena Sordo et al., "Prevalence of disease and age-related behavioural changes in cats: Past and present," *Veterinary Sciences* 7, no. 3 (2020): 85.

4   Lisa M. Freeman, "Double trouble: What's the best diet when your pet has more than one disease?," *Petfoodology* (blog), Cummings Veterinary Medical Center, 2020, vetnutrition.tufts.edu/2020/02/double-trouble-whats-the-best-diet-when-your-pet-has-more-than-one-disease.

5   Andre Tavares Somma et al., "Surveying veterinary ophthalmologists to assess the advice given to owners of pets with irreversible blindness," *Veterinary Record* 187, no. 4 (2020).

6   George M. Strain, "Hearing disorders in cats: Classification, pathology and diagnosis," *Journal of Feline Medicine and Surgery* 19, no. 3 (2017): 276–287.

7   Tammy Hunter and Cheryl Yuill, "What is cerebellar hypoplasia?," VCA Hospitals, n.d., vcahospitals.com/know-your-pet/cerebellar-hypoplasia-in-cats.

8   L.M. Forster et al., "Owners' observations of domestic cats after limb amputation," *Veterinary Record* 167, no. 19 (2010): 734–739.

Chapter 13: The End of Life

1   BBC News, "Oldest cat in the world, Scooter, dies age 30," 2016, bbc.co.uk/newsbeat/article/36292937/oldest-cat-in-the-world-scooter-dies-aged-30.

2   Dan G. O'Neill et al., "Longevity and mortality of cats attending primary care veterinary practices in England," *Journal of Feline Medicine and Surgery* 17, no. 2 (2015): 125–133.

3   Wei-Hsiang Huang et al., "A real-time reporting system of causes of death or reasons for euthanasia: A model for monitoring mortality in domesticated cats in Taiwan," *Preventive Veterinary Medicine* 137 (2017): 59–68.

4   Peter Sandøe, Clare Palmer, and Sandra Corr, "Human attachment to dogs and cats and its ethical implications," *22nd FECAVA Eurocongress, VÖK Jahrestagung 31ST VOEK Annual Meeting: Animal Welfare, Proceedings* 31 (2016): 11–14.

5   Naomi Harvey, "Imagining life without Dreamer," *Veterinary Record* 182 (2018): 299.

6   Sandra Barnard-Nguyen et al., "Pet loss and grief: Identifying at-risk pet owners during the euthanasia process," *Anthrozoös* 29, no. 3 (2016): 421–430.

7   Jessica K. Walker, Natalie K. Waran, and Clive J.C. Phillips, "Owners' perceptions of their animal's behavioural response to the loss of an animal companion," *Animals* 6, no. 11 (2016): 68.

8   P.A. Dingman et al., "Use of visual and permanent identification for pets by veterinary clinics," *Veterinary Journal* 201, no. 1 (2014): 46–50.

9   E. Weiss, M. Slater, and L. Lord, "Frequency of lost dogs and cats in the United States and the methods used to locate them," *Animals* 2, no. 2 (2012): 301–315.

10 Caroline Audrey Kerr et al., "Changes associated with improved outcomes for cats entering RSPCA Queensland shelters from 2011 to 2016," *Animals* 8, no. 6 (2018): 95; Jacquie Rand et al., "Strategies to reduce the euthanasia of impounded dogs and cats used by councils in Victoria, Australia," *Animals* 8, no. 7 (2018): 100.

11 L. Huang et al., "Search methods used to locate missing cats and locations where missing cats are found," *Animals* 8, no. 1 (2018): 5.

ACKNOWLEDGMENTS

THANK YOU TO everyone who has helped me with this book, and in particular the scientists, veterinarians, and animal shelter workers who have taken the time to answer my questions. Special thanks to Kristi Benson, Dr. Jill Bradshaw, Suzanne Bryner, Jean Donaldson, Bonnie Hartney, Stef Harvey, Kate LaSala, Kristin Lucey, Kim Monteith, Dr. Claudia Richter, Jessica Ring, Beth Sautins, Lisa Skavienski, Nickala Squire, Tim Steele, Dr. Rachel Szumel, Lyn Thomas, Roy and Frankie Todd, and Dr. Karen van Haaften. I'm grateful all the time for the ongoing support of everyone in the Academy writing group. My Ko-fi supporters have kept me going with kind words and coffees even when the going is tough. I'm also grateful to everyone who has supported *Companion Animal Psychology* with likes, shares, and words of encouragement over the years.

I am grateful to my amazing agent, Fiona Kenshole, for believing in me and for her incredible support. I'm also grateful to Trena White for helping me get started on all this.

A book is more of a team effort than most people realize, and I thank everyone at Greystone Books for their hard work to make this book a reality. In particular, I am grateful to my editor, Lucy Kenward, for her thoughtful comments and suggestions and her collegial approach to edits. I am indebted to Rowena Rae for her kindness, professionalism, and precision with copy edits. And a special thank you to Belle Wuthrich for another amazing cover design. Thanks too to Jennifer Croll, Megan Jones, Lara LeMoal, Kathy Nguyen, Hanna Nicholls, and Makenzie Pratt for their brilliant marketing and organizational support, and to Meg Yamamoto for careful proofreading.

I would also like to thank Jean Ballard and Fiona Kenshole for letting me use their gorgeous photos.

None of this would have been possible without Al. Thank you.

INDEX

Figures indicated by page numbers in italics

Also by Zazie Todd

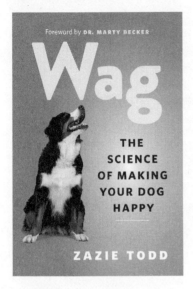

"The must-have guide to improving your dog's life."

MODERN DOG

Wag: The Science of Making Your Dog Happy

978-1-77164-379-5
304 pages • Paperback
$26.95 CAN • $19.95 U.S.• £16.99 U.K.

Winner, Best Book about Dog Behavior, Health or General Care, Dog Writers Association of America

What People Are Saying:

"If you've ever wondered what your pet is thinking and feeling, this book is a great place to start."

WUNDERDOG MAGAZINE

"Practical, compassionate, thorough—and based on science rather than wishful thinking—*Wag* is also a gift you should give to yourself and the dog or dogs in your life. I loved it!"

CAT WARREN, author of *What the Dog Knows*

"Well-written and packed with great advice, this book could fundamentally change the relationship between you and your dog."

DAVID GRIMM, author of *Citizen Canine*

"*Wag* will become a benchmark for the way that dogs need to be cared for: as sentient creatures experiencing emotions, just like us."

DR. PETE WEDDERBURN, veterinarian and author